"... An amazing accomplishment! After Darwin's *Origin of Species* and *Descent of Man* left his hands, they reoriented our thinking about the world and the place of human beings in it. But only a part of the message of those volumes became emblematic of the Darwinian revolution. David Loye now focuses attention on the other part of Darwin's message, which is one of cooperation, brotherhood, and a progressive understanding of the still developing nature of human beings. His trilogy, Darwin and the Battle for Human Survival, reveals what has been neglected with a passion and urgency not found in any other work. It is scholarship of deep humanity and needful wisdom, one that advances a new vision, but one thoroughly in the Darwinian spirit."

Robert J. Richards, national award-winning science historian and Darwinian scholar; professor, University of Chicago; author of the classic *Darwin and the Emergence of Evolutionary Theory of Mind and Behavior* and *The Meaning of Evolution*

"David Loye's is one of the few voices so desperately needed in the Darwin debates. Not only does he introduce "Eros" into the picture in a rational, sane, and supportable fashion, he makes the whole evolutionary theory hang together as "Eros-in-action," or, if you want, "Spirit-in-action" or "Love-in-action." It's been clear for quite some time now that the standard neoDarwinian synthesis can in no way account for the rise from dirt to Shakespeare, and new, believable theories are desperately needed. David Loye's is such a theory. The orthodox will cringe, and go on failing to explain evolution while bad-mouthing all other attempts, the Creative Intelligence folks will correctly spot all the missing holes in Darwinian theory and then—in an entirely unsupported move—fill in those holes by plugging them with Jahweh, a ridiculous move if ever there was one. The holes are supported by the data, Jahweh is not. David Loye's is both—spotting the holes, then filling them with something like a self-organization principle, which is a drive to higher levels of organization warranted by the data itself. Call that extra push "self-organization," or "Eros," or "Spirit," or "Love," or what you will, but it is fully justified by empirical data and scientific research. I know of no one doing this as thoroughly and carefully as David Loye. Read him, it's one of the most important topics alive today."

Ken Wilber, pioneering integral philosopher and psychologist; founder, The Integral Institute; author of *The Atman Project, Integral Psychology, Integral Spirituality,* and *A Brief Theory of Everything.*

"To shift from despair to hope as we face the renewed challenge of evolution: that and nothing less, is the challenge and the task taken up by David Loye in his profound, thorough, and deeply inspiring books on Darwin and evolution. Darwin has been misrepresented and misunderstood: a fate not unique among great scientists and prophets. Their insights are made to serve their followers' aspiration and confirm their followers' worldview—never mind what the thinkers and prophets themselves truly had in mind. When David Loye goes after what Darwin had in mind he is not only putting right the historical record; he is performing a crucial service in the cause of humankind. For today we, the species that calls itself homo sapiens sapiens, the knower-knower, faces its most rigorous test of intelligence: the test of viability. Can we, will we, survive on our home planet? We don't know yet, but what we do know is that the answer lies in our ability to discern the path that lies ahead for evolution. David Loye lights that path for us, and for that we owe him a profound depth of gratitude—one we can best repay by comprehending what he has discovered, and acting on it. Doing so is in our most vital and immediate interest."

Ervin Laszlo, pioneering systems philosopher, scientist, and global activist; editor *World Futures: The Journal of General Evolution;* founder, the General Evolution Research Group, the Club of Budapest, and Global Shift University; author of *Evolution: The General Theory, The Connectivity Hypothesis, You Can Change the World, and The Science of the Akashic Field*

"... congratulations on this stupendous output !!! If one has any doubt as to the correctness and importance of David Loye's reconstruction of Darwin's "lost theory," as laid out by him in half a dozen books over the last dozen years, this new trilogy should set those doubts to rest. In addition, clearly and entertainingly, he points the way to new paradigms for science that are crucial to our sustainable future."

Ralph Abraham, pioneering mathematician and chaos theorist; emeritus professor, UC-Santa Cruz, former faculty member UC-Berkeley, Columbia and Princeton Universities; author of *Dynamics: The Geometry of Behavior* and *Chaos, Gaia, and Eros*

"David Loye brings a brilliant career to its zenith in this trilogy on evolution. With the world facing threats like climate change, Loye's profound insights into the evolutionary advantages of both competition and cooperation form twin pillars supporting the move to a sustainable global order. Long overdue, this new work completes Darwin's work."

William E. Halal, leading futurist; professor, management science, George Washington University; president, TechCast LLC; author of *Technology's Promise*

"It seems to me David Loye is making a crucially important point in this book about politics and economics. Here we are, presently caught in a fierce struggle between one kind of politics and economics aimed at advancing human evolution while another kind of politics and economics is hell bent on driving us backward. Yet going by all the evolution theory that science and education presently focuses on you would never know this pivotal struggle for those who really care deeply and work to change things for the better in our world had anything to do with evolution. It looks like Loye's trilogy will at last hammer this crucial point across."

John Robbins, pioneering nutritional and social activist; recipient of the national Rachel Carson, Humanitarian, and Courage of Conscience Awards; author of *Diet for a New America, The Food Revolution,* and *The New Good Life*

"I read the whole book and love it! A must-read for all of us working for global transformation to a cleaner, greener, more equitable future for the human family. This crowning achievement takes David Loye's important work in earlier books to a new level of synthesis, clarity, and power. It provides the missing history and context we need to align and bring coherence to many of today's movements for positive futures."

Hazel Henderson, crusading moral economist, producer and host for the internationally distributed public television series Ethical Markets; author, *Creating Alternative Futures, Building a Win-Win World, Beyond Globalization,* and co-author, *Planetary Citizenship.*

"David Loye, as expected, does a brilliant job of deconstruction and reconstruction. He takes the conventional view of Darwin, and much like Hegel did of Marx, turns him upside down. We learn that it is not survival of the fittest that is the thrust of Darwin's work but moral sensibility. This re-reading of Darwin provides us with a theoretical framework for a new biopolitics and a new vision of the future. Written in an easy to read style, *Darwin's Second Revolution* is recommended for the scholar as well as the day to day facebook web surfer. Brilliant. And fun."

Sohail Inayatullah, pioneering futurist; professor, Tamkang University, Taiwan and the University of the Sunshine Coast, Australia; editor, *Youth Futures: Empirical Research and Transformative Visions, Macrohistory and Macrohistorians,* and *Chaos and Coherence in our Uncommon Futures*

"If humankind is to survive the 21st century, it will need to deal far more effectively with the escalating threat of nuclear warfare, ecological devastation, and the other threats to its existence. There is no better roadmap than David Loye's remarkable book *Darwin's Second Revolution.* Loye has brought to light a rich vein of gold still to be mined in Darwin's writings. We're all indebted to his wisdom, his scholarship, and his careful articulation of Darwin's revolutionary paradigms, needed more now than ever before."

Stanley Krippner, pioneering humanistic psychologist; professor, Saybrook University; former president humanistic psychology and psychological hypnosis divisions, American Psychological Association; author of *Human Possibilities* and co-author of *Healing States* and *The Realms of Healing*

"After a hundred years of evolution theory it's time to look at all of Darwin's writings—not just the ones encapsulated in the phrase "survival of the fittest." David Loye's account of contributions that revolutionize current evolutionary theory includes ignored ideas forwarded by Darwin himself, now fleshed out in avenues not explored in Darwin's time. An example is complexity theory, the finding of self organizing processes and how they apply to evolution by way of regulatory genes. All these new ways of approaching evolution do not as yet fit easily together, but the enterprise is young and Loye's contribution in calling attention to them will make it much easier to evaluate each and show how they can work together: a job Loye has heroically begun. This new way will shine light on our spiritual future, which rises out of a meaningful information revolution and leaves behind the now out-dated "material mankind" that was the heritage of the industrial revolution."

Karl Pribram, widely considered the greatest living brain scientist; developer holographic and holonomic brain theory; former professor Stanford, Radford, and Georgetown Universities; author of *Brain and Behavior, Languages of the Brain,* and coauthor *Plans and the Structure of Behavior*

"In this engaging book Professor Loye tears away the shrouds of secrecy and indifference that conceal the deep connections between a misunderstood Darwin and the financial and intellectual chaos that reigns in today's world."

Allan Combs, pioneering developer of consciousness studies; professor, California Institute of Integral Studies; author of *The Radiance of Being* and coauthor with Ervin Laszlo of *Thomas Berry: Dreamer of the Earth*

"Writing about love in a serious scientific context is a task reserved for thinkers who dwell on the sublime summits of human philosophy. By uncovering the long ignored emphasis on love in Darwin's original works, Loye reveals the deeply humane side to the evolutionary paradigm, which has been ethically degraded by the domination of concepts such as 'survival of the fittest' and 'the selfish gene.' Loye's writing is further a welcome combination of easy-flowing wording and solemn, old school elegance."

Vuk Uskokovic, nanotechnologist and systems philosopher, University of California, San Francisco, and President, UCSF Postdoctoral Scholars Association

"Loye's trilogy is a seminal step toward a critical pedagogy wedding bio-literacy and contemporary psycho-social scientific thought. Situated in transdisciplinary science, politics, and religion, Loye embraces controversy to radically expand basic Darwinian theory toward its yet unrealized true potential, as the action-oriented, functional scientific theory for the new millennium it was designed to be. This highly innovative interdisciplinary exploration of the span from biological to moral to spiritual evolution will offend many of our mono-disciplinary colleagues. The flow and scope surprised me, sparked my interest, and changed the course for my scientific thinking. It is a truism that the psychology behind the belief systems of adherents is at play in any enterprise, be it spiritual, material, religious, or scientific. Loye intrepidly explores this truism, bridging toward a transdisciplinary methodology that is our best, and perhaps only hope of efficiently and successfully sorting through future best-fit primary models of origin.

Christopher Peter Montoya, professor, psychology department, Thompson Rivers University, Canada; author of *The Migratory Theory of Genetic Fitness*

"David Loye is that rare scholar who investigates his subject beyond the normal parameters of science and history. His work on Darwin is brilliant and paradigm changing. For more than one hundred years Darwin's work has been mainly used to lock in a superficial belief in 'survival of the fittest". In this new book Loye shows the depth of Darwin's true thinking and the importance of this new perspective for today's scholars and general readers."

vi

William Gladstone, author of *The Twelve* and producer of the film Tapping the Source

"At the funeral of Karl Marx, Friedrich Engels said that, as Darwin had discovered the evolutional rules in biology, Karl Marx discovered the evolutional rules in human society. Since then human social practice has falsified Marx's revolutionary theory and human scientific practice has enriched Darwin's evolutionary theory. David Loye's work exploring the relation of social and cultural evolution as well as biological evolution to human practice over the past two centuries is a meaningful advance for Darwin's theory."

Min Jiayin, systems philosopher and research fellow, Institute of Philosophy, Chinese Academy of Social Sciences, Beijing; author of *Evolutionary Pluralism: A New System of Systems Philosophy*, and editor, *The Chalice and the Blade in Chinese Culture*

"*Darwin's Second Revolution* blows the lid off the Darwinian delusion that Darwin was as blindly mechanistic as some of his neo-Darwinian successors in evolution theory. This new story of Darwin comes at a time when the dominant narrative of mechanistic biology has hit the brick wall as an explanatory theory, being unable to account for a full range of evolutionary processes, including higher moral drives. Through carefully researched case studies, David Loye shows that Darwin was a genuinely humane, even spiritual human being. *Darwin's Second Revolution* shows that the dry rationality of contemporary mainstream science and academia have distorted the truth of one of the great men of history. Modern science depicts Darwin as a stubbornly courageous purveyor of truth, struggling to light the darkness of religious and social superstition. David Loye shows that the full spectrum of Darwin's light is yet to be made visible within dominant contemporary discourse. Loye's account of the real Darwin is a call for a deeper enlightenment, a powerful ray of hope."

Marcus T. Anthony, founder, Mind Futures (www.mindfutures.com), Hong Kong; author of *Integrated Intelligence, Sage of Synchronicity*, and *Extraordinary Mind: Integrated Intelligence and the Future*

Many scientists still regard NeoDarwinian evolution theory as the complete scientific account of the origin and future for human life, nothing essentially left unexplained. But through carefully analyzing what Darwin actually wrote, and its corroboration by hundreds of modern scientists, David Loye finds shocking evidence of a higher order, open ended evolution theory seeking to transcend and replace "survival of the fittest" Darwinism. Moral sensitivity far outweighs selfishness, mutuality is more meaningful than competition, love outpaces survival of the fittest a hundredfold. As Loye puts it, with striking new evidence of Darwin's intellectual support, we're driven to move beyond a science committed solely to the passive role of the so-called objective observer to the active role of science as *partisan* on behalf of and advocate of humanity. We need to focus on what a *full spectrum, action-oriented* theory of evolution should look like, and *how to actually build it.*

Pentti Malaska, pioneering futurist; professor of mathematics, Turku School of Economics, Turku, Finland; founder, Futures Research Center, and chairman, Academy for Futures Study Network

"David Loye 's trilogy, Darwin and the Battle for Human Survival, evinces an immense amount of research and apparent lifetime of passion for the deeper truths of evolution. Writing with erudition, Loye makes a convincing case that the moral sense is paramount in driving human evolution. I recommend this trilogy to anyone interested in the vital topic of evolution."

Steve McIntosh, Integral philosopher; author of *Integral Consciousness*, and the forthcoming book, *Evolution's Purpose*

David Loye writes with passion of the more cooperative and empathic humanity that emerges in Darwin's long ignored development of a theory of the evolution of moral sensitivity. He also writes eloquently of the role that a language and philosophy based on the use of self-organizational concepts of systems theory can provide in pursuit of an evolutionary cultural trajectory to establish such a global society. He writes of the pioneering of Kurt Lewin, the great American psychologist and expatriate from the Frankfort School of Critical Theory founded to apply social science to issues of domination. Among those working today to transcend simple reductionist evolutionary theory using complexity theory, he writes of his partner, cultural evolution theorist Riane Eisler, my brother Ralph Abraham and Stuart Kauffman, the great evolutionary biophysicist.

Frederick David Abraham, co-founder and past president, Society for Chaos Theory in Psychology and the Life Sciences; former research professor, UCLA, UC-Irvine, University of Vermont, and Silliman University, Phillipines; author *A Visual Introduction to Dynamical Systems Theory for Psychology*, and editor *Chaos Theory*

An astonishing work! Loye's passion for the essence of Darwin renders a great service. *Darwin's Second Revolution* brings us face to face with the very nature of being human—the caring, social, species-with-a-conscience mammal that evolution made us, the qualities that so moved Darwin. Loye's engaging voice makes this a page turner that gives students, educators, scientists and policy makers a vivid history lesson and much more. We discover the true spirit of Darwin, brimming with the moral sense of our species. This book is powered by love, insights and wisdom for our age and beyond. Loye shows us that in our battle for survival the regressive blinders can and must give way to the progressive urge in every human, in every culture. Our future depends on it.

Raffi Cavoukian,
Singer, author, founder of the Centre for Child Honouring

"David Loye has opened the minds and hearts of this whole generation to the real meaning of the world-changing work of Charles Darwin. Loye's work is vital to us all."

Barbara Marx Hubbard, evolutionary visionary activist; President, Foundation for Conscious Evolution; author of *Conscious Evolution* and *Emergence*

"Charles Darwin's Vision of Hope, Reborn."

Kenneth Bausch, renaissance scholar; Executive Director, Institute for 21st Century Agoras; author of *The Emerging Consensus in Social Systems Theory* and *Body Wisdom: Faith in Chaos*

"Darwin's Second Revolution is absolutely brilliant; one of the most important contributions in print about how we can re-understand human nature and re-invent a viable human society."

Thom Hartmann, crusading journalist and host, national radio Thom Hartmann Program; author of *Threshold: The Crisis of Western Culture, Unequal Protection: The Rise of Corporate Dominance and the Theft of Human Rights,* and *Rebooting the American Dream*

ix

Darwin's Second Revolution

By the author

The Healing of a Nation
The Leadership Passion
The Knowable Future
The Sphinx and the Rainbow
The Partnership Way (with Riane Eisler)
An Arrow Through Chaos
Darwin's Lost Theory of Love
The Story of a Family
The Evolutionary Outrider, Editor
The Great Adventure, Editor
Bankrolling Evolution
Measuring Evolution
Brave Laughter
Return to Amalfi
Tangled Tales of the Book Trade
3,000 Years of Love
100 Days of Love
1001 Days of Love
Darwin on Love
Darwin's Second Revolution
The Battle of the Books

DARWIN'S SECOND REVOLUTION

*Book I: Darwin and the Battle for
Human Survival*

DAVID LOYE

BENJAMIN FRANKLIN PRESS

Published by
The Benjamin Franklin Press
First Edition

1. Science. 2. Biography. 3. Evolution. 4. Religion. 5. Philosophy

Cover by John Mason. Production: Cassandra Gallup-Bridge.
Back cover photo: Don Eddy.

For more information about Benjamin Franklin Press books
www.benjaminfranklinpress.com
elliotsanders@benjaminfranklinpress.com
The Benjamin Franklin Press, P.O. Box 51936, Pacific Grove, CA 93950
Phone: 831-626-1004. Fax: 831-626-3734

Dedicated to all those who
first glimpsed the ghost at the birthday party,
all those who for a century fought for Darwin's second revolution
against the counter-revolution,
and to Riane

CONTENTS

PROLOGUE
THE GHOST AT THE BIRTHDAY PARTY

Every year we're reminded that Charles Darwin was born on February 12, 1809. As with the tick of a celestial clock, this becomes the date for celebrations throughout the world of a great annual birthday party. For a brief time the old pictures and stories are taken out of the mothballs in the trunk of the global attic. We see Darwin's ship, his beard, his dog, his home, his greenhouse with orchids, his 374-year-old sea turtle. Early in the 1990s, however, I stumbled across something akin to the shock for many a murder mystery. For amid the cutting of the cake and the blowing of horns, I first glimpsed the ghost at the birthday party.

There was this sense of something off kilter. Of a strange discrepancy between the theory and the man. Could something very big, I wondered, have been left out of what we've been taught about Darwin and his theory of evolution?

Having gained my credentials, prestigious faculty posts, and publication of influential books,[1] I decided to apply my training as a scientist to an investigation. As reported in the book that serves as a grounding data base and launching point for this book and trilogy, *Darwin's Lost Theory*, I had an electronic copy of *The Descent of Man* that made possible a computerized word search.

Where should I begin? The answer for nine people out of ten if you asked what is Darwin's theory of evolution seemed logical. So into the FIND slot went the time-worn phrase "survival of the fittest": in a split second came the first shock.

Only twice in that book of 475 fine print pages does this prevailing identity for Darwin's theory of evolution appear—one of the times being Darwin's apology for ever using the term survival of the fittest.

Should we try a polar opposite?

Into the FIND slot went the word love and out came the jolt that cracked open the way to the lost completion for his theory. For in *The Descent of Man*—in which Darwin tells us he will now move on from the study of *prehuman* evolution he wrote about in *Origin of Species* to what advances *human* evolution[2]—he wrote *95 times* about love.

Could this be? I checked the Index. Now the crack widened into a gap that soon became mind-boggling. After 100 years in every edition of *Descent*, in all the main languages for our species throughout our whole world, as of the 200th anniversary celebration of Darwin's birthday, in 2009, there was still only a single entry in the index for love!

One entry for love—versus 95 times in the text.

By now it was obvious something of possibly great importance lay ahead.

What about the new tag for Darwinism in our time. What about "selfish genes"?

In book after book sociobiologists, evolutionary psychologists, and hordes of eager interpreters have told us that selfishness is the prime driver for human evolution.

The favorite way of demonstrating its power has been to show that it lies at the heart of our foolish illusion that altruism, or caring for others, is anything more than just what's in it for me in the end.

"Selfishness," I found the ghost of Darwin thundering out of the yellowed pages, is a "base principle," which accounts for the "low morality of savages."[3]

What could be the polar opposite of "selfishness?"

Of *moral sensitivity* I found that in *Descent* Darwin wrote 92 times—versus only 6 entries in the Index.

Of competition, 12 times; of cooperation—called mutuality or mutual aid in Darwin's time—27 times.

And then this. For much of a century we were taught that according to Darwin evolution is some mighty mass of biological processes automatically, unwittingly, shaping us to some unknown end. Yet in *Descent* I found he wrote 200 times of the active, self-organizing, values-driven mind and brain.

Then came the shock of shocks. Not buried in some obscure place easy to miss. On the next to the very last page, in the section of *Descent* clearly labeled *Concluding Remarks,* there rises from the page this passage.

Important as the struggle for existence has been and even still is, yet as far as the highest part of our nature is concerned there are other agencies more important. For the moral qualities are advanced either directly or indirectly much more through the efforts of habit, by our reasoning powers, by instruction, by religion, etc., than through natural selection."[4]

Could this be possible? How and why had this been ignored? For if this was true if this was what Darwin had actually found and really believed—I saw what for a century we tamely accepted as the prevailing story and structure for much of science and society crumbling before me.

One thing I must make absolutely clear here. This is not a case of replacing Darwin's theory of natural selection, as outlined in *Origin of Species,* with the "lost," "higher" theory of *The Descent of Man.* They are the two halves of a single over-riding theory embracing the whole of life. The problem is what was ignored, distorted, or most importantly left out by

3

Darwin's successors. Given the uncovering of his intended *completion* for his theory, one can see clear through to the core sickness in our national and global mind space.

One can see how environmental destruction, nuclear overkill, terrorism, galloping corruption, rule by corporation, and on and on, driven by the mindset of "survival of the fittest" and "selfish gene" Darwinism, are like metastases that fan out from an immense cancer at the core..

Certainly other factors were involved, but had we had the completed theory—and thereby completed *story,* inspiring aspiration rather than excusing degradation—would we have had to endure a "century of almost unbroken war," in which 187 million of us were killed?[5] Would Hitler have been able to rally the most scientifically advanced nation at that time behind the Nazi atrocity and Holocaust? Would we have had to quiver under the threat of atomic annihilation during the forty year Cold War?

And would we have entered the 21st century with the worst president in American history—riding on the mad multibillion dollar political and economic roller coaster of elections and lives governed by the greed of the few versus the need of the many? With progressive vision swept aside by regressive surreality? With rightists and creationists dancing with glee over success for the politics of lunacy?

Indeed, as we'll see in Part II, an alternative title for this book might well have been *Darwin and the Politics of Lunacy.*[6]

What happened? What can we do about it now?

Far too large and important a question to be covered in a single book, *Darwin's Second Revolution* is the first book in a trilogy written to uncover the answer. The result, on one level, is a new story of the development of evolution theory throughout the 20th century. But on a higher level of meaning, spanning past, to present, to future across centuries, it is something more. For I believe that out of this new unfolding of the story emerge the missing pieces to the puzzle of what must be done to shift from despair to hope for the 21st century.

This trilogy is about the aftermath to Darwin's first revolution, which over 100 years ago laid the foundation for his theory of evolution. It's about the rise of a scientific counter-revolution, which on being seized by regressive politics and economics as well as regressive religion drove—and still drives— human evolution off track. But now we're moving into a third stage that ends the mystery and opens the way to survival. We've moving into the hope for our children and our children's children of Darwin' second revolution.

Driving uphill against the stranglehold of what century after century has held us back, this trilogy is about the across-the-board thrust within progressive science of modern studies that update and expand Darwin's long ignored, moral action-oriented completion of theory. By revisiting who we were told we were, but now know we actually are—and even more importantly, who we can and should become— we'll uncover the scientific corroboration of Darwin's actual insistence that moral sensitivity, education, and love, rather than "survival of the fittest" and "selfish genes," are the primary drivers of human evolution.

All in all, it's apparent we've gone off track in evolution, time is running out for us, and the question is how do we get back on track.

On one hand are those of us who—blind captives of an antihuman paradigm and the counter-revolution of lower expectation—automatically, tragically, work to check us in place or drive us backward.

On the other are gathered those of us who, out of at least 100,000 years of yearning, generation after generation have worked to advance evolution—that is, to build the better world through the revolution of *higher* expectation.

This was the challenge for the 20th and now the 21st century that began with the agony of Romanes.

PART I

THE TRIUMPH OF THE NEOS

ONE
THE AGONY OF ROMANES

"Why?"

Up from the yellowed pages, over the confusion and bloodshed of a whole century, long forgotten but now of urgent meaning for our time, still rises Romanes' dying cry.

Rich, widely respected, known throughout the science of the time as Darwin's disciple and a foremost British psychologist, George Romanes was the founder of the field he called comparative psychology. In keeping with Darwin's own strong interest, considerable research and thought, Romanes' goal for this new field was to probe the similarities between the minds of animals and humans.

Typical among Romanes' pioneering books were *Mental Evolution Among Animals,* and then *Mental Evolution in Man.* Of special interest today, however, is a book that quickly dropped out of science and history. In keeping with the mysterious disappearance of Darwin's own completion of theory, following publication in 1897, Romanes' *Darwin and After Darwin* virtually dropped out of print until picked up and published online in its entirety first by Project Gutenberg in 2008 and then more recently by Google Books.

Almost wholly ignored for a century, today it looms as a prophecy of haunting accuracy.

Why, Romanes wrote, "not only do the Neo-Darwinians strain the teachings of Darwin; they positively reverse those teachings— representing as anti-Darwinian the whole of one side of Darwin's system ..."[1]

Why, he asked, was one of the new Neo-Darwinians "unjustifiably throwing over his own opinions the authority of Darwin's name."[2]

More specifically, why "so greatly have some of the Neo-Darwinians misunderstood the teachings of Darwin, that they represent as 'Darwinian heresy' any suggestions in the way of factors 'supplementary to,' or 'co-operative with' natural selection."[3]

The great man himself, Romanes wrote with difficulty but determination as he lay there dying, "was always ready to entertain 'additional suggestions' regarding the causes of organic evolution—several of which, indeed, he himself supplied."[4]

What was being done to Darwin was, for Romanes, like a knife being twisted in the fatal wound.

Of the "new writings ... now habitually represented by the Neo-Darwinians as setting forth the views of Darwin in their 'pure' form," he wrote in sorrow, anger, and despair that "both in conversation and in the press, we habitually meet with complete inversions of the truth."[5]

Being manufactured to replace the living Darwin he had known, Romanes charged, was a new scientific pseudo-religion and a new breed of pseudo-priests to exploit the dead. Being enshrined was a "scientific creed ... not a whit less dogmatic and intolerant than was the more theological one which it has supplanted ... and while it usually incorporates the main elements of Darwin's teaching, it still more usually comprises gross perversions of their consequences."[6]

Out beyond the bedroom window to which his eye strayed as he lay dying, was the world of such increasingly poignant meaning—this world to which what he felt, and wrote, and prayed for time to complete and publish, was a matter of complete indifference.

1894 it was, with things going on as usual. The British and Belgian governments were signing a secret accord dividing up all the people, gold, diamonds, and whatever else there was of value in Central Africa between them. London Tower Bridge was opening. George Bernard Shaw's play "Arms and the Man" premiered in London, and Sherlock Holmes "Adventure of the Empty House" was published. In America, Coxey's Army of the unemployed sets out from Massillon, Ohio, for Washington, D.C. In New York City 12,000 tailors go on strike protesting sweat shops. The country is electrified by Edison's kinetoscope for moving pictures, Elwood Haynes successfully tests the new automobile at 6 miles per hour, Indian chiefs from the Sioux and Onondaga tribes meet to urge their people to renounce Christianity, and 136,000 mine workers go on strike in Ohio for a pay increase.

Elsewhere, Japan is defeating China in the Battle of Ping Yang. 6,000 Armenians are being massacred by the Turks in Kurdistan. Debussy's ballet "L'apres-midi d'un faune" premieres in Paris. A vaccine for diphtheria is announced in France by Dr. Roux. A first commentary on evolution in comic book form, "Origin of a New Species," is launched by Richard Outcault, soon to become world famous as the creator of "Buster Brown."

And Romanes lies here dying of what—after year after year of a painful decline—the doctors finally decided was a fatal brain tumor.

Many times his thoughts would have gone back to that day when Darwin and he first met. Himself this mere tyro, this awkward nobody who'd written a letter of admiration, and after a brief correspondence received a letter in return from Darwin to come visit him. How the great man had rushed forward to seize his hands with delight upon his arrival. "How glad I am that you are so young!" Darwin had exclaimed.[7]

During Darwin's final decade theirs had been a unique relationship. Young George Romanes had become not only his worshiping disciple, but in regard to where Darwin's mind was going, his closest intellectual companion.

Their relationship "reached an intensity that seemed to have no rival," Darwinian authority and historian of science Robert J. Richards tells us. "Their frequent meetings and correspondence bespoke the insinuating bonds of father and son. When Darwin died in 1982, Romanes grieved as he had previously done for no man."[8]

"Half the interest of my life seems to have gone when I cannot look forward any more to his dear voice of welcome, or to the letters that were my greatest happiness. For now there is no one to venerate, no one to work for, or to think about while working," Romanes wrote to Darwin's son Francis.[9]

The rest of Darwin's designations in his will went elsewhere, but to Romanes Darwin left all his notes on his pioneering exploration of the field of psychology.

As first his eyes, then his balance, then ultimately his legs and even speech declined and left him, Romanes had labored to write his tribute to Darwin.

Whether "the misrepresentation be due to any unfavourable bias against one side of his teaching, or to sheer carelessness in the reading of his books," what was happening was both inexcusable and reprehensible. The Neo-Darwinians, he wrote—for it was Romanes who first coined the phrase—had set out to "positively reverse" Darwin's teachings.[10]

But already, like the ghost of Hamlet's father, Romanes' was a voice trying to work upon the conscience of the living from the grave. Before the book's completion, he died.

"I myself believe that Darwin's judgement with regard to all these points will eventually prove more sound and accurate than that of any of the recent would-be improvers upon his system," Romanes wrote as the room, and everything in it, and the years with Darwin, the adventure of science, the treasured cameraderie, the thrill of debate, the glow of praise, the clasp of love, Scotland, the touch of children, and the world on and on beyond all that steadily dissolved into a blur.[11]

Here is the quote he used, from Darwin himself, to underline his own astonishment and concern.

"But as my conclusions have lately been much misrepresented, and it has been stated that I attribute the modification of species exclusively to natural selection," Darwin had written as far back as in *Origin of Species* originally, "I may be permitted to remark that in the first edition of this work, and subsequently, I placed in a most conspicuous position–namely, at the close of the Introduction—the following words: 'I am convinced that natural selection has been the main, but not the exclusive means of modification.'"[12]

Then came the lament of Darwin that became Romanes's own.

"This has been of no avail. Great is the power of steady misrepresentation," Darwin had written. "But the history of science shows that fortunately this power does not long endure."[13]

It was a fine sentiment, a comforting hope to take the edge off Darwin's lament. But having seen "misrepresentation" steadily grow rather than decline during the decade after his mentor's death, Romanes had not been optimistic.

For as if he spoke with the supposed power of a reincarnated Nostradamus, the story thereafter of what both he and Darwin feared did for a very long time endure.

TWO
DARWIN'S LOST THEORY

To go the the heart of the matter in as few words as possible, I will briefly condense what I found lay behind Romanes' concern. Here is the core of what is in page after page of Darwin's long ignored own writings. As I report at length in *Darwin's Lost Theory*, here is the key to the story over a century of what happened, what could have happened but didn't, and what must happen now.

To begin with, I found the obvious. In *The Descent of Man* Darwin continues to write of the fundamental and by now well-established scientific evidence for the impact of natural selection on variation in shaping evolution. I also found his firm discernment of the drive of selfishness. No doubt about it. But now—as we've seen in the startling quote I've cited—Darwin went on to insist that "other agencies" become of greater importance at our species' level of evolutionary emergence.

Which is the "right" theory?" Which best fits who we were, are, and can become?

Contrary to the conclusion we customarily leap to, this is not a matter of two theories, one good and one bad. Digging and testing findings for more than a decade, I found that back of the word counts and quotes, Darwin outlines a compelling, carefully reasoned *moral* action-oriented completion for a single theory and story of evolution.[1]

In other words, instead of what he intended, what happened is a matter of fixation on the beginning of a single over-riding theory and story and lock-step forgetting and ignoring the end.

The central point for the Darwin we've been denied for over 100 years, as Romanes decried, is that the sole driver for *human* evolution was and is *not* natural selection, or "survival of the fittest," as popularized. Natural selection readied the stage for our emergence. In the theatrical sense it dressed and coached us for our early roles. But then, like Shakespeare, we graduated from being only the actor in a crude and bloody play written by others to being the writer, actor, and producer of a wondrous new variety of plays of a higher and progressively liberating order.

Moreover, central to this higher order for Darwin was, and is, our capacity for the "moral sense"—that is, moral sensitivity, an evolutionary inbuilt thrust within us for the development of a pivotal sense of right versus wrong.[2]

Add love, education, and progressive religion and philosophy to the mix for development of an inborn drive of moral sensitivity and you get the lost power of Darwin's completion of his theory—an advance now confirmed by carbon-dating, fossil and DNA analysis, and other amazing tools of modern paleontology, as well as by the brain research of MacLean and many others.[3]

Further clarifying the built-in power of what Darwin and prior philosophers called the moral sense, in Darwin's completion of his theory are two parallel tracks. One is confirmed by paleontology and biology. The other is confirmed by psychology and other social sciences. Both in the end are confirmed by brain research and the new field of evolutionary systems science.[4]

The First Track

For the lost Darwin, foreshadowing the *second* revolution, first came what the poetry of centuries, modern advertising, and Freud made obvious,

15

which biology and paleontology confirm: the primordial emergence among organisms of what he called the *sexual* instinct.

Fossils show this happened about a billion years ago with the emergence of meiotic sex, or the mating of two organisms for reproduction.[5] For Darwin the explosive step up for the sexual instinct next led over time to the emergence of a *parental* instinct.

Here again his pioneering insight is confirmed by paleontology and brain research, which shows the capacity for caring for—rather than eating—one's offspring did come next with the arrival of the early reptiles, about 250 million years ago.[6]

The next step for Darwin led beyond the fierce boundary of self and offspring to emergence of the capacity to care for others. Fundamental to his theory thereafter, this was the portentous arrival of the *social* instinct. Here again, both paleontology and brain research confirm his insight, dating the arrival of the mammals to about 216 million years ago.[7]

Last then for Darwin came the emergence of the capacity for *emotion* and finally *reason,* with a notable flowering with the emergence of our species confirmed for somewhere between 100,000 and 200,000 years ago.[8]

Defying the stereotype of Darwin as the mournful old guy with the long beard, confirming the genius of Darwin at a remarkably early age, this burst of step by step insight was first recorded in the long unpublished private notebooks into which, at age 28, Darwin poured everything exploding within his mind on his return to England from the voyage of the Beagle.

Level by level precisely confirmed by the brain research of Paul MacLean and others, this is the step by step upward evolutionary track from the hypothalamus, to the lower limbic system, to the higher limbic system, to the prefrontal cortex in the brain of every one of us.[9] This is the sequence to which, in order to complete his theory of evolution, Darwin returned in *The Descent of Man*.

This is the portentous sequence for his development of the evolution of the moral sense—*not survival of the fittest*—as the prime driver for human evolution.

The Second Track

In a parallel analysis, Darwin develops his case for how beyond biological evolution our cultural evolution unfolded. In keeping with the earlier implanting of our sexual, parental, and social instincts, through cultural evolution came the widening impact of our *caring* for others.

Given our capacity for emotion and reason, this led to *reflection* on the consequences of one's behavior.

That, in turn, led to the development of *language* to share and compare insights.

Then, with the global spread of the capacity for language—and the sharing of what seemed to better rather than worsen our situation on this planet—came the mind-binding impact of *habit*.

Through doing the same thing over and over again, through repetition and refinement generation after generation over centuries, the guidance of right versus wrong embodied in customs, norms, rules, values, and morals was—and is—cemented in place.[10]

Stage by stage, this lost Darwinian conclusion is supported by modern brain research, cultural anthropology, linguistics, and the psychology of learning and developmental and social psychology.[11]

Here we can begin to see how the completed theory we have been denied for over a century consists of two halves.

Just as with the building of a house, the rearing of a child, or the move from grade to high school and college, there is a foundation, primarily established by the interaction of natural selection and variation, the domain of natural science. This Darwin wrote of in *Origin of Species. But then comes the completing superstructure*—the domain of social and systems science, of which he wrote in his early notebooks and in *The Descent of Man.*

Most vital for our bloody, rudderless, and often despairing time, in projecting the second half, Darwin shows how, through a process of both personal and mass cultural maturation, the thrust of the *moral sense* unfolds.

For an example applied to many other fields today, here we can see how one hundred years earlier Darwin foreshadowed the thrust of Abraham Maslow's revolutionary development of humanistic psychology. We can see the prefiguring of the still widening human potentials and integral philosophy movements.[12]

In the lost Darwin I also found what we'll return to in chapters seven and eight. For here, over 100 years ago, were insights anticipating what are still leading edge aspects for science and society as we move into the maelstrom of the 21st century.

Chaos, Complexity, and Self-Organizing Theory

A concept that defied the dumbing down of the late 20rth century to seize the popular mind was the "butterfly effect" of chaos theory. This is the idea that even a seemingly insignificant thing, like the stamp of a butterfly, can, through systems interconnections, influence change thousands of miles away.

Actually first starting in *Origin of Species*, then accelerating in *The Descent of Man,* Darwin repeatedly called attention to the fact that even a small change in one part can produce considerable change in other parts of a living organism.[13]

Whether operating on systems as large as global weather patterns or as small as the interaction of neurons in the brain, modern studies reveal the principle is the same.[14]

Indeed, Darwin apparently ran up against so solid a wall of indifference that in succeeding editions of *Descent* his calls for attention to what he called *correlated variation* become plaintive, even pleading.

And so we come to perhaps the greatest excitement for late 20th and early 21st century science: the implications for all fields of science of self-organizing theory.

This is the most dramatic case I know of for how long it can take for science to catch up to reality. In personal and political terms, the root idea is that we don't have to put up with a raw deal, whether handed down from on top by "God," natural selection, or by our "fellow man." We can rebel. Thanks to evolution, we have a voice in shaping our own destiny.[15]

And what did I find in Darwin? As detailed in *Darwin's Lost Theory*[16], in his analysis of the process from thought to action within the mind of a bee, Darwin carefully describes the steps whereby we and all other organisms have a voice in shaping our destiny.

As Darwin never named it, I have suggested the term *organic selection* for this pioneering observation.[17] At least eleven times in *Descent* he develops this insight into the operation of self-organizing processes—again only to be ignored for over 100 years.

Darwin and Religion

Now we move beyond science to the religious side of the battle for human survival. For contrary to the stereotype, as we've seen in the key quote in the prologue, Darwin was *not* the enemy of religion.

Contrary to the Satanic anti-Christ of the creationists and the scientific half-truth for many of their opponents, I found that a reading of what Darwin actually wrote reveals that throughout his life there was a running battle, or love-hate fest, in his mind toward religion and spirituality.

Clearly for him the ignorance, the violence, and the lunacy of regressive religion was an abomination. But science has been quick to forget that Darwin was originally formally schooled at Cambridge not for science, but for the ministry. Also generally overlooked is the number of ministers who were close lifelong friends and his ferocious alignment to the slave-hating Unitarian church of the Darwin and Wedgewood family's faith.[18]

As for the idea of God, while rejecting it for himself, he endorsed it as a needed source of comfort and inspiration for others. In *Descent,* for example, he writes of "the ennobling belief in God."[19]

Given that we must make up and catch up for the loss of a century, inevitably this summary must raise questions. The answers to most of them can be found in the copiously indexed mother text of *Darwin's Lost Theory,* with more to be answered in the next books of this trilogy.

Our task now is to move on to the immense question this loss raises of how and why we got so far off track in evolution, and how to get back on track.

As with psychotherapy, or simply over our years of growing up, we advance by coming to see who we were, now are, and can become. Likewise, given the key to past, present, and future of Darwin's "lost theory," we have the chance to re-examine how we got here and can now move forward.

We'll now return to Romanes' time—before he died, still living.

We'll return to the beginning of the heroic, tragic, myopic, ironic, and in the end inspiring story of the struggle between the revolutionists and the counter-revolutionists—and what for our century has become the battle for human survival.

THREE
WEISMANN'S RATS AND WALLACE'S SPOOKS

In Romanes' time much was made of his battles with two scientists he railed against as the worst offenders among those he was first to call the Neo-Darwinians.

The most obdurate and menacing of these opponents was Frederich Leopold August Weismann, a German biologist, whose face as he aged became so dour and forbidding as to look like he ate nails along with his sausages and *Brochen* for breakfast.

Weismann first made his mark on history with what became known as the Weismann Barrier. This was his theory of the impossibility for evolution to occur in any other way except through biology shaped by the action of natural selection on variation.

Amid the confusion of everything going this way and that at the time, Weismann's firm Prussian assertion of the sacred Barrier was widely welcomed. For some time thereafter, it helped wall off further straying in the direction of anything more than natural selection being involved. It further seemed to provide the *coup de grace* to a worrisome open question for many at the time.

The problem was what to do with the variation for natural selection to work on. In other words, for natural selection to drop in from "above" to pick the winners from the losers among the varying individuals in all species, there had to be some explanation of how all of the variations came to be for natural selection to select.

It was evident there had to be something else at work transmitting change from generation to generation.

For lack of any other explanation they could agree on, the classic theory, which most rejected as far too simplistic, was the pre-Darwinian assertion by Jean-Baptiste Lamarck that evolution was passed on from one generation to the next by by "use-inheritance."

A favorite example was the case of the long neck of the giraffe. To Lamarck and a good many others it looked like this must over time had been caused by the creature that became the giraffe having to stretch its neck to reach the tender leaves of the trees it loved to dine on.

Bang down came the gavel of Weismann's Barrier and out went Lamarck, presumably forever. To cement the sacred Barrier firmly in place—to prove there could be no possible way for evolution to occur other than biologically, untouched or in any way altered by life experience —Weismann first made the case for an involvement of "germ cells" with "somatic cells."

While this was a mystery to all but the most up-to-date of biologists, it seemed to make good sense to many of the good minds of the time. To seal the case for his theory experimentally, Weismann then, one after another, generation after generation, cut off the tails of 1500 rats.[1]

Off with tails for the parents, who then produced children with tails; off with their tails, who then produced children with tails; and so on for 1500 of them.

Over all that time no rats without tails were produced; thereby presumably ruling out any possibility for change to occur except via the track for "pure" Darwinian biology. Case closed—except for obstreperous dissidents like Romanes.

The difficulty for Romanes was that this conclusion not only ran up against Darwin's own repeated insistence that, while natural selection was firmly, unquestionably the main operator, there was more to it. Weismann's rats and Barrier also bumped squarely against a personal agenda for Romanes, to which we'll come.

Romanes'other much-publicized opponent was none other than Darwin's prestigious co-discoverer of the principle of natural selection, Alfred Russel Wallace.

Physically, Wallace was a striking figure. Even in a time of magnificent masses of face hair up, down, and sideways, his adornment was a standout. Mentally, he was also commanding. In papers and in commentary on the intricacies of the evolution argument in his time, he was up to speed with, and if so impelled ready to jump ahead from the rest.

While respecting Wallace's intellect, what galled Romanes was that here again was someone who—against Darwin and his own agenda —was insisting that within the material or natural world only the principle of natural selection prevailed.

In the midst of the wild flurry of charge and counter-charge, in *Darwin and After Darwin* Romanes suggested the following as a way to "easily escape this confusion."

All one had to do was to "remember that wherever in the writings of these naturalists there occur such phrases as 'pure Darwinism' we are to understand pure *Wallaceism,* or the pure theory of natural selection to the exclusion of any supplementary theory."[2]

But now there entered a new wrinkle in a wobbly paradigm. For while Wallace was obdurate about natural selection as the one and only solution for the material world, he had decided to split evolution into movement in a supernatural as well as natural world.

"My opinions on the subject [of man] have been modified solely by the consideration of a series of remarkable phenomena, physical and mental, which I have now had every opportunity of fully testing," Wallace

23

had in 1869 earlier warned Darwin. They "demonstrate the existence of forces and influences not yet recognized by science."[3]

One demonstration he had in mind, now famous in the annals of spiritualism, was what a doubter might well call the Case of the Floating Fat Lady.

Carefully guarding in many ways against the possibility of fraud, in 1866 Wallace had gathered together in his own home a group of friends and the medium Agnes Guppy. They sat together around a table holding hands. Then in some way none could explain afterward, with no sensation of Mrs. Guppy's hands having parted with those on either side of her, the others gathered there saw the inconceivable. Noiselessly, all 300 pounds of her, was suddenly sitting in a chair that also suddenly appeared on the table before them.

Another time, in the dead of winter, fresh flowers normally available only in springtime, appeared.[4]

"All were absolutely fresh as if just gathered from a conservatory," Wallace reports. "They were covered with a fine cold dew. Not a petal was crumpled or broken, not the most delicate point or pinnule of the ferns was out of place."

Among reports of scores of similar happenings in his books on paranormal phenomena, Wallace tells of another Guppy sitting during which a friend of his asked for a sunflower. Promptly, so the story goes, a sunflower six feet long fell on the table with a large mass of earth dangling about its roots, as if it had just been uprooted.

Such things were widely mocked as the trickery and fraud of charlatans. But Wallace was joined in his time by a large number of prestigious and notably hard-headed physicists and psychologists, who stood together to certify the truth of comparable experiences. Among them were Darwin's brilliant cousin and pioneering experimental psychologist

24

Francis Galton, pioneering psychologists G.T. Fechner in Germany and William James in America, pioneering physicist Oliver Lodge, and the chemist and editor of the *Quarterly Journal of Science* William Crookes.[5]

From years of exploring and puzzling over the meaning of such inexplicable things, Wallace concluded that "Man is a duality, consisting of an organized spiritual form, evolved coincidentally with and permeating the physical body, and having corresponding organs and developments."[6]

This spiritual twin of a wider contact and greater knowledge, he contended, acting as guide and guardian, looks after us throughout our lives, with the cumulative result that these spiritual twins act as outriders that outrun natural selection to lead our species into the future.[7]

As if this weren't enough to add further problems to the mix, in a world soon to be split into a bloody battle between capitalism versus communism, Wallace was also that suspect thing for both sides: a fervent socialist and passionate advocate for peace.

So Weismann and Wallace presented to Romanes what he saw as an intolerable affront to Darwin's and his own research and beliefs. In both loomed the threat to the liberating and open-minded venturing of Darwin of a potentially disastrous closing in of the mind to fixate on natural selection alone.

Barely underway, were we to be shut off from further growth in any other direction in science?

In the case of both Weismann and Wallace, were we to have the door not only slammed in the face on sexual selection and community or group selection, both of which Darwin insisted were vital complementary theories for evolution. Were we also to be denied all of what, out of the intimacy of his years with Darwin, Romanes already intuited and the science of our time supports of the "lost" half of Darwin's theory.

And what if Wallace and all the other scientific eminences exploring the world of spooks were successful in convincing people the supernatural was still involved?

Was the whole matter to be thrown back into the clutch of a religion that Darwin and Romanes and so many others had fought so hard for so many years to overcome?

To cap it all, for Romanes there was a personal agenda. There was his desire to find a place in history for his own pet theory of "physiological selection."

This was a complicated matter of the varying fertility and inter-crossing of species, of which I—as well as apparently droves since then—was forced to conclude it was something that only a mother or misguided graduate student in biology could love.

It was a heavy load for a man with a powerful sense of historical continuity and responsibility. One can't help feeling there must have been a connection between Romanes' frustration and the even greater frustration of the monstrous turn of events whereby he was to be taken off before his time by a brain tumor.

So where are we?

Can we at this point in our story talk of revolutionaries and counter-revolutionaries? No. So far we have only a heady mix of super bright revolutionaries. For all the talk of "battle" and "enemies" that back then made for sparkle in the journals, professional gossip, and a beginning for wider press coverage as the topic of evolution heated up, we only have a mutually respectful group of revolutionaries arguing among themselves over what constitutes their revolution and how and where it should be headed.

At this point, however, we may detect something else beginning to nudge at the mind. Something to which 100 years of history and human experience and the updating of sociology, political science, and evolutionary systems science has become sensitive. Something to which in times of transition from revolution to counter-revolution one becomes super-sensitive. One might say it is a sense of a hidden third party to what up to now has been a friendly argument.

It is almost as if within the twist and turn of scholarly argument there is an over-riding, hovering, invisible something that seems to slide in at every useful juncture to coo or whisper, "Let's you and him fight. Let's build this into a battle to the death."

FOUR
THE MINDFUL MEASURE OF MORGAN, BALDWIN, OSBORN, AND WILLIAM JAMES

Darwin, as we've noted, left to Romanes his extensive notes on the development of the field of psychology. It seems clear this was to encourage Romanes to push the study of evolution on into the realm of social and systems science as his scientific heir. Now once again, in designating his own heir to finish *Darwin and After Darwin* after his death, Romanes left to a younger admirer the responsibility to do the same for him.

Conway Lloyd Morgan was not a colorful figure with a dramatic story, as Darwin or Romanes had been. Coming from South Africa, however, with a degree in metallurgy and mining rather than either biology or social science, through self-education, with the aid of Romanes, he rapidly became a leading British psychologist.[1]

What knowingly linked him to Romanes—and unknowingly to Darwin—in the development of a theory of evolution was his interest in departing the ever more well-traveled super highway of biology for the relatively undeveloped upper trail of psychology. Like Darwin in his final years, and Romanes from the beginning, Morgan made his mark with the study of the mind. Like Romanes, he sought the grounding for development of an understanding of intelligence first in animals, then in humans.

He first became known for Morgan's Canon. For a while this, ironically, was to become more of a shackle than a liberator to the study of the human mind.

"In no case may we interpret an action as the outcome of the exercise of a higher mental faculty, if it can be interpreted as the exercise of one which stands lower in the psychological scale," Morgan's Canon decreed.[2]

In other words, you couldn't say a dog was smart, or loving, or even possibly had a sense of humor, if it could be shown that this was just your interpretation of behavior that actually had a subhuman cause.

As time went by, however, Morgan departed this Canon. He shifted to what first became known as "organic selection," and then "The Baldwin Effect."[3]

Organic selection, Morgan felt, could be the explanation for Lamarck's conviction that change within a lifetime could be passed on to the next generation. The answer, he felt, might lie first in psychology, and only thereafter in biology.

What was this and how did it work?

The customary explanation is so involved that all but the career-minded soon go blank and drop away. The contrasting connotation in terms for *organic* versus *natural* selection suggests the difference. On being confronted with change, the organisms that prosper are those with the *mental* ability to adapt to changed conditions. This smarter, hence successful, remnant intermarries, breeds, multiplies, and so over time what was at first a learned response becomes embedded as a biological instinct.

Fired up with this possible answer to much that still mystified everybody else, Morgan departed England for a lecture tour in America. In Boston he stayed for a week with William James, to whom we must guess he imparted his idea. Then in Chicago, in a lecture, he first unleashed it publicly.

To his surprise, the next speaker, the American psychologist James Mark Baldwin, announced essentially the same seemingly independent discovery. Moreover, a few months later American anthropologist Henry Fairfield Osborn, also apparently independently, announced roughly the same discovery.

Except for the usual skeptics, it seemed a new case of the idea whose time had come. Then, in wholly unexpected ways, back into the picture came Weismann and Wallace.

This was the Weismann of the Weisman Barrier barring the door to any involvement of human intelligence in evolution toward which Morgan was now headed. Yet now to Weismann Morgan credited this first step toward a viable wedding of psychology and biology.

The original insight, Morgan magnanimously ventured, was an idea of Weismann's of something he called "intra-selection" at work.[4]

To cap it all, a statement by Romanes of almost exactly the same idea had appeared in his posthumous *Darwin and After Darwin*, which had been finished and edited by Morgan.[5]

"...if functionally produced changes, and changes produced in adaptive response to the environment, are ever transmitted in a cumulative manner, a time must sooner or later arrive when they will reach a selective value in the struggle for existence," the dying Romanes had written.[6] And what was this exciting new breakthrough, after all?

No doubt exacting scholars in this area will disagree. But what I see from the viewpoint of where we are in science today are two things. First is that essentially this venture was the next step up from the notion of "self-adaptation" that a "Reverend Henslow" was earlier trying to peddle uphill—with first discouragement and then grudging interest from Romanes in his final days.[7]

Secondly is that Morgan, Baldwin, Osborn, and earlier Romanes and the mysterious Reverend Henslow were all driven by a split-level vision.

On the lower, conscious level, was the desire to solve a problem in their varying interpretations Darwin's *Origin of Species* theory. They were driven to find out how they might be able to squeeze a squeak of freedom of choice into an otherwise hapless organism being pounded into shape by the force of something incredibly larger and more powerful than itself.

In the surrealistic turn this tale so often takes, it was as if natural selection had become merely a change of name for God the all powerful, whose every command must unquestioningly be obeyed.

Although psychologists, they were still by training not only ruled but frequently over-ruled by the paradigm for biology. Still imprisoned within the data of their rats—and later crabs, sea worms, spiders, fruit flies and vampire bats—they were trying to solve the problem of something larger that widened out into the world in all directions from the laboratory and the tidy field trip.

On one hand was the pull of the reality of the human world, to which Darwin had said he was turning in *The Descent of Man*—which became the reason for his intense interest in psychology in his final years; which thereafter inspired Romanes, then Morgan; which already was exploding into the spread for all the fields of social science destined to expand in significance during the 20th century.

On the other hand were the constraints of a magnificent theory hobbled within natural science and a paradigm for the prehuman world.

They were, in short, trying to solve the larger real and pressing problems of the present and future human world within the constraints of a theory of evolution imprisoned within the pre-human world and the deep past.

The Mysterious William James

But what of William James in this picture? Throughout the brief bubble of what became known as The Baldwin Effect—Baldwin having adroitly managed to label "organic selection" with his own name—James remains a shadow in the background in most accounts.

This seems strange. In his understanding of the mind as well as in his eloquence as a writer and in general influence, James towers above all the others. Here was the pioneering philosopher and psychologist, whose book *Principles of Psychology* became the most popular textbook in the history of American psychology, sparking advance in practically every field. Why then does he seem to have remained such a silent partner in the ancient fight to establish "organic selection" (i.e., the Baldwin Effect) in science?

I think the answer lies in the fact that out of a crisis in his life so intense he contemplated suicide, James wrote "My first act of free will shall be to believe in free will."[8] Further in the same vein he wrote that mind is preeminently a "fighter for ends."[9] This is the William James of whom an astute scholar of his work wrote: "The doctrine of freedom meant that mind was not identical with brain, nor its slave."[10] And again, hearkening to the central issue for the lost Darwin, this is the William James who wrote: "If this be a moral world, there are cases in which any indecision about its being so must be death to the soul."[11]

I think he avoided much active engagement in The Baldwin Effect controversy because his eye was on something much bigger. More surely than Romanes or any of the others, broadened with the scope and understanding of a major philosopher, I believe his intuitively was the vision that was originally Darwin's.[12]

It was the vision that by the end of the 20th century had become the explosion of excitement over the drive of *self-organizing theory* within chaos and complexity theory. Characterized by James as "a fighter for ends," it was the drive of mind that became the late 20th century spark of autocatalysis for Ilya Prigogine, autogenesis for Vilmos Csanyi, autopoesis

32

for Humberto Maturana and Francisco Varela, and sparked my own small contribution in moving chaos theory from natural into social science.[13]

In any event, this early thrust for what became the track for revolutionaries was doomed to an abrupt end. A police raid on a black brothel in Baltimore uncovered evidence that one of the white men they picked up attempting to use a false name was none other than James Mark Baldwin, prominent John Hopkins University professor.[14]

Below the headlines the story gradually exploded throughout the South and the scientific community. Here was a major pioneer in the study of moral development—for as such, long afterward, he came to be recognized[15]—caught with his pants down, as they say, involved with a prostitute of the very worst color for the time.

Baldwin fled to Mexico and then to Paris. There he was welcomed with, one gathers, only a chuckle, and continued to gain scientific honors. But in America not only was his name mud. Ironically, in the way systems dynamics operate, for a time the scandal seemed to blank out everything else associated with his name.

The point of the "butterfly effect" of chaos theory is that what is at first a small "attractor" in one part of a vast system can set off waves, which through widening resonation, can take over and change the system as a whole. There was more to it, but in retrospect one can see this at work in what next happened.

As scandal swept the Baldwin Effect and all the work associated with it off the stage, this left the play for evolution theory in the hands of those who were now becoming the counter-revolutionaries. For now the way was clear for the rise of full-blown Neo-Darwinians—with one notable exception.

FIVE

KROPOTKIN, LONESOME PRINCE
AND HAPPY REVOLUTIONARY

Of those who during the early years continued to push for the larger vision and completed theory that Darwin's meant to leave, the most engaging was Peter Alexeyevich Kropotkin. A Russian Prince by birth, Kropotkin abandoned status, wealth, and home to become a political and scientific revolutionary.

On the surface Darwin and Kropotkin were poles apart politically and economically. Darwin was a wealthy, canny investor in capitalism. Kropotkin the anarchist was rock firm against both capitalism and authoritarian communism. But on the subject of moral sensitivity as a primary driver for evolution they were fully in tune.

"Nature," Kropotkin wrote shortly after the time we've looked at so far,[1] "was represented by the Darwinists as an immense battlefield upon which one sees nothing but an incessant struggle for life and an extermination of the weak ones by the strongest, the swiftest, and the cunningest: evil was the only lesson which man could get from Nature."[2]

But this, he hammered across, is only the mainstream interpretation of *Origin of Species*, upon which both Darwin's successors then and the 20th century almost as a whole became fixated.

Of the Darwin of *Descent* and the second and completing half for his theory, Kropotkin wrote:

"There is, he [Darwin] showed, in Nature itself, another set of facts, parallel to those of mutual struggle, but having a quite different meaning:

34

the facts of mutual support within the species, which are even more important than the former, on account of their significance for the welfare of the species and its maintenance."[3]

This "extremely important idea," Kropotkin wrote, is the key to Darwin to which "most Darwinists refuse to pay attention ... which Alfred Russel Wallace even denies."[4]

It is, he tells us, to this neglected aspect of both Darwin's vision and his theory that, he, Peter Kropotkin—by picking up and expanding the term and concept of *mutual aid* first used by Darwin—has set out to give his life. This is the motivation behind his writing *Mutual Aid*, *Ethics*, and scores of articles and lectures he gave in exile in Switzerland, France, England, America, as well as finally, after the revolution, back in Russia, before the Bolsheviks took over.

It is an uncanny experience to read Kropotkin from where we are today at quite possibly the most important juncture in the history of our species on this planet. For Kropotkin's two great books, *Mutual Aid: A Factor of Evolution*, and *Ethics*, read as if they were the voice of a reincarnated Darwin. But here, at that time roughly four decades after Darwin's death, it is the voice of a Darwin given a new life to which was added what Kropotkin enjoyed—better health, more wide-ranging scholarship, freedom from the weight of all the ins and outs of biological theory that Darwin was forced to juggle with, and above all, more adventure, which at times Darwin had rather poignantly yearned for.

Within all of his life Darwin had only the voyage of the Beagle—that single supreme five years of freedom to explore the earth and life beyond his home, his social class, and the sleepy little village of Down in which he lived out his days. But Kropotkin's was a life of incredible richness.

David Loye

Memoirs of a Revolutionist

His life comes alive in one of the most engaging of autobiographies, Kropotkin's *Memoirs of a Revolutionist*. His childhood as a page in the court of the Czar Alexander. His early years as a explorer and natural scientist mapping Asia in a new way. His early life as a prince partying with the royalty at night, preaching revolution to the serfs in the day. His imprisonment in the grim Fortress of Peter and Paul. His dramatic escape to support himself, as Karl Marx did, as a journalist writing reports and commentaries on world events for European and American newspapers. His closing eminence as the foremost philosopher of anarchism in his time.

Anarchy still remains mainly associated in the American mind with the idea of potential assassins and bewhiskered bomb throwers. Out of this mindset the feisty Emma Goldman was deported to Russia and Sacco and Vanzetti were executed in 1927.

The root word and concept, however, comes from the Greek *anarchos*, meaning "no rulers." As for Kropotkin, most anarchists abhorred bloodshed and were devoted to peace. Their heresy was to be believe there should be no organized government, that people should be left to work things out for themselves without centralized control.

I personally find the idea immensely impractical, but understandable within the context of the history out of which it emerged.

Following posthumous publication in 1925, Kropotkin's great book, *Ethics: Origin and Development*, in keeping with the pattern that is becoming familiar, was mainly out of print until quite recently. Yet in all of the 20th century I found this to be the book providing the most useful review of the advance of moral thought from the most ancient into modern times.

Here's a typical passage from the opening pages.

When we see that scores of thousands of different aquatic birds come in big flocks from the far South for nesting on the ledges of the "bird mountains" on the shore of the Arctic Ocean, and live here without fighting for the best positions; that several flocks of pelicans will live by the side of one another on the sea-shore, while each flock keeps to its assigned fishing ground; and that thousands of species of birds and mammals come in some way without fighting to a certain arrangement concerning their feeding areas, their nesting places, their night quarters, and their hunting grounds; or when we see that a young bird which has stolen some straw from another bird's nest is attacked by all the birds of the same colony, we catch on the spot the very origin and growth of the sense of equality and justice in animal societies. And finally, in proportion as we advance in every class of animals towards the higher representatives of that class (the ants, the wasps, and the bees amongst the insects, the crane and the parrots amongst the birds, the higher ruminants, the apes, and then man amongst the mammals), we find that the identification of the individual with the interests of his group, and eventually even self-sacrifice for it grow, in proportion.

In this circumstance we cannot but see the indication of the natural origin not only of the rudiment of ethics, but also of the higher ethical feelings.[5]

As the 20th century wore on, this was the kind of science—and talent for making it come alive as a writer, as had been true of Darwin—increasingly written off as "just anecdotal." Where were the statistics, the mathematics, the bare bones battling of the abstractions of theory?

In all of my search, in all of the 20th century, I found no other scientist or scholar who so clearly saw and reported two of the most important foundations Darwin was trying to establish for our successful understanding of, and alignment to, evolution in *The Descent of Man*.

One is that Darwin was outlining the completion for his theory of evolution—that the structure was of a first half in *Origin*, and a second and completing half in *Descent*.

The other is equally fundamental, but even more widely unknown. For beginning with *Origin of Species*, Darwin was making the case for not one, but the *two* parallel systems for the course of evolution. This I write of in "Two Stream Versus Single Stream Theory and Consequences" in ending Reflections for Book II: *The Battle of the Books*.

Going light years beyond the prevailing picture for mainstream science, this is an insight going at least as far back as the concept of Love versus Strife for the colorful early Greek philosopher Empedocles.[6] It resurfaces in modern times as Eros versus Thanatos for Freud; in negentropy versus entropy in physics; in anabolic versus catabolic processes for Walter Cannon and others in endocrinology.[7] Moreover, bursting out of the cloister of science into the turbulent evolutionary reality of politics and the question of the direction for human evolution, there is this: As I develop in *Bankrolling Evolution*, further develop here in chapter ten, and will return to in Book II, the two stream versus single stream perception resurfaces in liberal versus conservative in terms of their functions in evolution.[8]

This elusive but vital point about Darwin's theory was quite clear to Kropotkin.

"There is, he showed, in Nature itself," Kropotkin wrote, "another set of facts, parallel to those of mutual struggle, but having a quite different meaning: the facts of mutual support within the species, which are even more important than the former, on account of their significance for the welfare of the species and its maintenance."[9]

This insight further accounts for Kropotkin's claim that this "explains also the simultaneous existence of two different codes of morality among civilized nations."[10]

I explore this insight and the following observation for Kropotkin at length in my forthcoming book *The Glacier and the Flame II: Redefining Evil.*

"Moral science appears, therefore, as the search for a compromise between a code of enmity and a code of amity—between equality and inequality," Kropotkin wrote.

"And as there is no way out of that conflict—because the coming of the industrial state will only be possible after the cessation of its conflict with the militant state—nothing can be done for the time being save to introduce into human relations a certain amount of 'benevolence' which can alleviate somewhat the modern systems based on individualistic principles."[11]

This is something to think about, as we no longer confront the problem of a conflict between what we know today as the industrial state and the militant state. Rather, increasingly we seem to face the horrifying prospect of a melding of the two into a single global devastator.

And so, after the entries and the exits for this early cast of characters, we come to the end of Act One. Now the action passes to Act Two.

SIX

INCORPORATION OF
THE NEO-DARWINIAN MONOPOLY

A shy, retiring monk is at work in the garden of a monastery in Europe. We move in for a close up. The face is bluff, clean-shaven, black hair neatly combed, with the firm jaw of a man of undeviating determination, eye glasses sparkling with the light of early morning.

His gaze concentrated on the odd task at hand, he is clipping the anthers from the bloom of scores of blue, white, and multi-colored sweet peas on vines wrapped around the supporting poles that surround him. He next dusts the stigma for the bloom with some pollen from a bottle in his cassock. Then he puts a small sack over the bloom, gently tying it in place.

This is Gregor Mendel, the Austrian monk for whom not just half his contribution, as was the fate for Darwin, but all of it was essentially lost, and then after his death dramatically rediscovered, changing everything.[1]

Following Weismann and Wallace, the concentration for the new evolutionists pretty well generally settled on natural selection as the big, over-riding, master principle for evolution. The theory was still under fire not only from the Church, however, but from scientific holdouts. Contrary to the impression that acceptance came quickly, it would be a half century before natural selection gained reasonably wide acceptance.

The delay came because of how the action had turned to the lesser of the great pair—the squirming Variation of all the organisms little and big, which Natural Selection was supposedly out there somewhere in biological space hovering to select from.

40

One of the most persistent investigators nibbling at the question of what might be the most viable theory was Gregor Mendel. Others were at work cross-breeding plants to look for changes from parent to child. But he was the sole Sherlock or Hercule for biology's mystery story. Over seven years he cross-bred nearly 30,000 sweet peas while keeping the careful records that helped establish the new science of genetics.

For thirty four more years others would wrestle over the question of how the changes of variation were passed on from generation to generation for natural selection to select from. But early on Mendel had much of what his successors came to agree was the answer. In the relatively inflexible ratios for dominant, regressive, and hybrid strains for white and red varieties of sweet peas, his patient labor and careful records began to reveal the pattern for the transmission of traits for all organisms: plants, animals, and ultimately ourselves.

He privately published his findings, sent copies to leading scientists of the time, and was mostly ignored by all but one of them. The contribution of this eminence, Swiss botanist Karl Wilhelm von Nageli, was to suggest that Mendel might do better by switching from sweet peas to another plant. This sent poor Mendel off into a biological blind alley so discouraging that he decided to give in to pressure to become Abbot of the monastery with no time for science any more. He died, however, confident that his was a major discovery, the importance of which would be recognized some day.

"Some day" became thirty four years later. In a replay of the simultaneous Darwin-Wallace discovery of natural selection, Mendel's work was independently rediscovered by three scientists, in Holland, Germany, and Austria. Soon thereafter the incorporation of the Neo-Darwinian monopoly that Romanes had decried began to take hold.

The Neo-Darwinian synthesis was unquestionably not only a major achievement for 20th century science. It still accounts for advances in vast areas of our lives, ranging from the quantity and quality of the food we eat to the wide range of drugs and medicines for medical treatment—both major factors in why many of us live so much longer today than earlier.

At the same time, however, mindful of Romanes' lament, it could be said some very large chickens came home to roost.

The Battling Bateson

A crucial role in the acceptance of innovations is that of the brash champion for the innovator. Darwin's champion had been his formidable "bulldog" T.H.Huxley. For Mendel a posthumous champion emerged out of a battle between two biologists who were at first friends, and then became the bitterest of enemies.

Mendel's bulldog was the biologist William Bateson. Father of today's better known son, anthropologist Gregory Bateson, Bateson senior began his study of evolution with an exotic marine worm known as *Balanoglossus*. On shifting to shellfish within salty ponds in the Central Asian steppes, he reached the decision he was to literally worry like a bulldog for years to come. Darwin's belief had been in evolution as a slow, gradual process. Bateson, however, decided on a radical break. Evolution moved ahead, he claimed, with "saltations."[2]

One might think this referred to the salty ponds for Bateson's shellfish. This might have been the original impulse, but soon throughout the field of biology this became a fighting word with the insider power of derivation from the Latin—"saltus,"not for salt, but meaning "to leap." Thereafter, the literature is saturated with "saltation" and "saltatory" for what one would normally call a jump ahead or break in process.

Saltation also occasionally became the word for what we know today as "mutations" and later "punctuated equilibrium."[3]

The friend who became an enemy was Walter Frank Raphael Weldon. Rising rapidly to eminence at Cambridge University, Weldon began as predominantly a crab man, but soon abandoned crabs for a passion for the then new field of statistics. Here, it seemed obvious to him, was the answer to the questions of evolution at all levels.

Here was this field slipping and sliding all over the place for lack of a sufficiently sophisticated level of measurement. And here was statistics, like the arrival of the Lone Ranger on the scene.

Moreover, behind the rise of the statistics that entranced Weldon lay the amazing story of Darwin's brilliant cousin, Francis Galton. A child prodigy who was reading by the age of two, who at age five knew some Greek, Latin, and long division, by age six Galton had moved on to Shakespeare for pleasure, and poetry, which he quoted at length.

Galton went on to become the most ingenious pioneer in the development of measures and methodologies for psychology and the social sciences more generally. Correlation, normal distribution, standard deviation, regression toward the mean, regression analysis, the first use of questionnaires and surveys—all these hefty statistical tools for research were Galton's invention.[4]

Within Galton's genius for measurement, Weldon saw the potential for using the new statistics to track change from generation to generation among species. Not only did the solution to the mystery of variations seem at hand. The heady thought of making one's mark by toppling the master beckoned. For it seemed to Weldon that he might be able to find and show that the hypothesized over-riding power of natural selection could be explained by nothing more than quirks of process at the supposedly lowly variation end.

Moving to Oxford University, he teamed up with pioneering biometrician Karl Pearson. Joined by Galton himself, out of a Royal

Society committee the three issued a historic statement.

"The questions raised by the Darwinian hypothesis are purely statistical," they proclaimed, "and the statistical method is the only one at present obvious by which that hypothesis can be experimentally checked."[5]

Out of throwing down this gauntlet came Weldon's collision with the battling Bateson. For not only was Weldon a gradual process man. He took out after Bateson's pet idea of saltation and the saltationists.

To further rub salt in the wound, as Bateson claimed Mendel's work was evidence of saltation, Weldon began to pooh pooh the significance of Mendel.

Bateson and Weldon were both extremely bright, eloquent, and—even for science—especially vituperative advocates for their positions. Bateson raged against the Weldonites "perverse inference, " "slovenly argument," "misuse of authorities, reiterated and grotesque." He accused them of not "acting in good faith as genuine seekers for truth."[6] Weldon died in 1906, but the argument raged on between Bateson and Weldon's ally, the even more vituperative Karl Pearson.

Meanwhile, others were quietly raising what now seems the obvious question: Why couldn't it be both? Why couldn't it be gradual process with also now and then jumps?

Morgan and the Sexual Economics of Fruit Flies

A jumps man who was also skeptical about both Darwin's principle of natural selection and Mendel's ratios for the transmission of traits for sweet peas now enters the picture. Thomas Hunt Morgan began his search for answers at the lowly level for variation with sea spiders. Then—in keeping with the long established "baby step" level for Darwin with barnacles 100 years earlier—he moved on to sea acorns, ascidian worms, frogs, sea urchins, fish, and earth worms. Then with a move to Columbia University, he graduated to the work with fruit flies that made him famous worldwide.[7]

Of all the creatures studied by the workers and management for the steady spread of the Neo-Darwinian monopoly, *Drosophila Melanogaster* was to prove the most productive of choices.

"It could be bred by the thousands in milk bottles," Darwin biographer Ronald Clark notes. "It was easy to find, needed little space, cost little and lived on simple food." Most importantly, where with humans you had to wait up to twenty years for a parent to produce a suitable child for experiment, a *Drosophila* egg "hatched, turned into an adult, and was itself producing more eggs within ten days."[8]

Starting in 1908, in the 16 by 23 foot space of his famous "fly room," Morgan supervised the mating of thousands of *Drosophila*—estimates range from 10,000 to 30,000.

For two years nothing notable happened. Then early in 1910 came the first experimental production of a bona fide mutant—a white-eyed fly after two discouraging years of the mating and birth of nothing but red-eyed flies. Thereafter, the sex life of the fly room came to rival the tabloids and Hollywood scandals, as Morgan's fundamental contributions to science spread from the scientific journals out through the popular media internationally.

At last the function of the chromosome and the gene in the transmission of variation became clear. Particularly compelling was Morgan's description of how in the flow of chromosomes genes were both linked to one another and crossed over in the matings to produce variants within species. Indeed, so gripping was Morgan's description of sex as the selector that it threatened to unseat Darwin's natural selection as the prime driver.

"Nature makes new species outright," Morgan wrote. "From this point of view, the process of evolution appears in a more kindly light than when we imagine that success is only attained through the destruction of all rivals ... Evolution is not a war of all against all, but it is largely a creation of new types for the unoccupied, or poorly occupied places in nature."[9]

This concept of variations seeking a niche to fill, I feel compelled to remark, was another one of many observations reaching into the future that I found in the lost Darwin, which I'm currently pursuing.[10]

Fisher, the Power of Numbers, and the Rise of Eugenics

The next player to stroll upon our stage was the notably handsome, near-sighted, and for a time massively under-employed Ronald Aylmer Fisher.[11]

Though like Galton early displaying a remarkable aptitude for mathematics, young Fisher was blocked from getting anywhere that mattered by World War I. Bad eyesight prevented him from joining the British army. He lacked credentials for anything else that mattered, so he settled for clerical work and small teaching posts. Only after the war did he gain peripheral entry into the world of science as a journalist. It was during this period, however, although lacking the credentials to be recognized, he wrote a piece later rediscovered and acclaimed as a major work.

What at last looked like Fisher's great opportunity to be taken seriously came when he submitted an article to biometric mogul Karl Pearson for approval. After delay, in a brief dismissive note, Pearson finally replied, "I do not think in the present state of affairs that the paper is wide enough to be of much interest from the biometric standpoint."[12]

Shortly afterward, in 1918, urged on by Darwin's son Leonard, the Eugenics Education Society sponsored publication through the Royal Society of Edinburgh; and at last Fisher was underway.

For in the paper dismissed by Pearson, Fisher provided the scientific jump from subhuman to human evolution. By applying his genius for mathematics, Fisher showed how both the biometric approach to large populations and the results of Mendel's experiments could be synthesized and applied to directing the evolution not just of fruit flies, nor just to the development of better strains of corn or herds of beef. Now it looked like the road was clear for the use of evolution theory in experiments with the most difficult but greatest of animals: ourselves.

This leap ahead confronted Fisher, all other evolution theorists, and indeed all of us today with the ancient moral and political question that now became the dilemma for science: how might one develop and apply a new science of evolution to breed better humans.

Of deep, abiding, and in our time now of critical evolutionary importance, it shoved a psychologically, politically, and economically naive group of biologists out into the portentous overlap between the field of science and the realities—and what could all too easily become the horrors—of politics, economics, and morality.

The Eruption of Eugenics

Politically, Fisher was an ardent conservative. Behind him lay the traditional conservative alignment to heredity. A mere 150 years earlier the divine right of kings had prevailed as the supreme example of a belief in a fundamental split within humanity between privileged and non-privileged blood lines.

Along with this belief had come the enduring split into higher and lower classes, favored and disfavored races, and, most fundamentally, male domination and female subordination.

For the burgeoning Neo-Darwinians, all this matter of classes, politics,

47

and dominators and dominatees was, of course, outside the proper realm of science.

It was that messy world one dealt with in one's own private ways. Inevitably tainted by a non-scientific "subjectivity," it was this periodically troublesome matter one had to consider when going for grants from foundations or governments.

But outside the purview of the new science of evolution they were building, the onrushing sweep of history was inescapable.

With the French Enlightenment and American revolution had come the shift away from heredity to environment as the liberal and progressive answer to the question of how we are to shape the better human.

"Give me your tired, your poor, your huddled masses yearning to breathe free" was emblazoned on the base of the Statue of Liberty.

But now, with the shift from a focus on prehuman to human evolution, the Neos were confronting the world with hard new evidence that heredity was a primary, if not even *the* primary consideration for the advance of human evolution.

What the new science revealed— which a majority of scientists were slow to comprehend and respond to—was a monumental shift in responsibility. From the religious point of view, no longer could human evolution be left up to God. Nor from the scientific point of view, given the new understanding of the power of the dynamics of variation, could the world be left to drift willy nilly with no guidance other than what a hypothetically blind Natural Selection provided. Both the nature and the desired direction for our species and planet called for Artificial Selection on the grand scale.

It was clearly up to science and enlightened political leadership to more directly and effectively intervene in evolutionary process.

Out of the challenge now raised by Fisher's work[13]—how might one best develop and apply a new science of evolution to breed better humans—erupted the burgeoning eugenics movement.

Most powerful was the alignment of Francis Galton, founder of eugenics, of British philosopher and pioneering systems scientist Herbert Spencer, Karl Pearson, and among many other famous names for the time H.G.Wells, George Bernard Shaw, and even Sidney Webb, founder of the ultra-progressive, socialist Fabian Society.

Among biologists against the eugenics movement was Morgan. But the chief opposition came from progressive American sociologists, such as Lester Ward and Charles Cooley—both of whom saw in eugenics a spin off from the pernicious doctrine of Social Darwinism and survival of the fittest.[14]

And so we come to the final scene for Act II. The question of how far and how fast they could or should go with eugenics now faced the Neo-Darwinians. Skirmishes on this and purely scientific matters continued. Now and then, however, came a lull permeated with a sense of expectancy.

Now with a roll of the drums from the media there arrived on the scene a band of both familiar and unfamiliar players headed by the man of the hour: biologist Julian Huxley.

Huxley, Dobzhansky, and the Great Synthesis

Grandson of Darwin's famous bulldog T.H.Huxley, gifted as a writer, an excellent synthesizer of the work of others, Julian Huxley sounded out the principal biologists of his time, and in 1942 published *Evolution: The Modern Synthesis.*[15]

Here at last, in this massive 600 page work, divisions were bridged and for a time many arguments were ended. For here was a decisive wedding of the still hotly debated Darwin-Wallace principle of natural selection with the explosive, contentious probe of the nature of variation following rediscovery of Mendel's launch for the impending reign of the gene.

Specifically, the new synthesis consisted of a consensus among the principal players. As summarized by biographer Ronald Clark, this binding historic agreement was that "gradual evolution could be explained in terms of mutations and their recombination, which produced genetic variation worked on by the process of natural selection; and that evolutionary phenomena, including the macroevolutionary processes and speciation revealed by paleontology, could be explained in terms of known genetic mechanisms."[16]

Within the incorporation of what by now had become a Neo-Darwinian monopoly, you might say that for a while Julian Huxley served as the chairman of the board. Among board members were the man of the hour for the eugenics movement, Ronald Fisher, and key contributors to evolution theory J.B.S. Haldane, Sewell Wright, E.B. Ford, Ernst Mayr, Bernhard Rensch, Sergei Chetverikov, George Gaylord Simpson, and G. Ledyard Stebbins.

In a special relation to Huxley, however, was the biologist who in a sense served as co-chairman, the Russian born geneticist and evolutionary biologist Theodosius Dobzhansky.

Theirs was an enormous accomplishment. Sending waves of immense influence out through all the fields that determine what is and what isn't to be included in our minds, the Neo-Darwinian synthesis deserved every bit of the celebration it's received as the most important achievement in the science of living systems during the 20th century.

All seemed to be for the best in "the best of all possible worlds" —except for two problems, which gradually expanded into the battle for control of 20th century and now 21st century mind.

Within science the problem was by and large a gentlemanly and gentlewomanly argument between those steeped in a theory of evolution oriented to the past versus those working to build a theory not only oriented to, but useful for, shaping the future.

Within the mind space for biology, and more generally natural science, this was a battle that to the non-involved outsider often seemed little more than a distant tempest in a teapot.

But outside the board room of Neo-Darwinism, Inc., hammering on the door, or refusing to wait any longer and storming out, were unhappy share holders in the destiny of Planet Earth. Primarily they were social scientists. But also many others, bewildered and frustrated, both secular and religious, clamored for change.

It was proclaimed a battle of "reductionists" versus "expansionists." At the core this was a struggle between those who claimed the secret to evolution could be found and solved within the life and territory of the creatures and processes of the microscope and laboratory, and those who claimed the exploration and understanding of evolution must radically expand to embrace a much wider territory.

It was, one might say with progressive bite, an argument between proponents of a theory of subhuman, inhuman, and inhumane evolution versus the proponents of a theory of *human* evolution. For the politically sensitive the problem was a theory all too much about adaptation and accommodation, all too little about assertion and aspiration. For the economically sensitive the problem was a theory cocked to the greed of the few rather than the need of the many. Or as a street fighter might put it: A theory of knuckling under to the Boss and the Fat Cats versus a theory of standing up to fight for one's rights.

But out beyond the mind space of science and theory pro and con loomed the exponentially larger problem of a reality radically accelerated by a science of consequences now beginning to run wild, out of control.

In the name of Darwin, Mendel, and the seemingly iron-clad doctrine of eugenics, long priding itself as being the home of the brave and the free, the United States became the first country in the world to enact compulsory sterilization for the mentally retarded and mentally ill, also the deaf, blind, epileptic, and the physically deformed. Native Americans, as well as African-American women, were sterilized against their will in many states, often without their knowledge while they were in a hospital for other reasons, for example, childbirth. To curb the supposed transmission from parent to child of possible "criminal" genes, for a while it also became a popular treatment for prison inmates.

Prior to 1931, over 65,000 "defectives" were sterilized in 33 American states under state compulsory sterilization programs, with laws to this effect remaining on the books as late as 1956.[17]

Now, within the overlap between the mind space of biology and the mind space of politics, economics, and religion, began what by the 21st century was to loom as the environmental end game for our species and possibly our planet.

On one side, based on the co-optation and degradation of the first Darwinian revolution trumpeted by Social Darwinism, were the advocates and exploiters of a theory of survival of the fittest and selfishness *uber alles* as the driver for human evolution.

On the other side—keying without knowing it to Darwin's "lost" completing half—was the thrust of the second Darwinian revolution and the increasingly distressed advocates of a theory of love and moral sensitivity as the driver of human evolution.

Out of the ensuing confusion and mess of science, politics, economics, religion, and morality rose Adolf Hitler and the Nazis. But also out of the shock rose the desire of both Neos and anti-Neos for something radically better attuned to the building of a better world.

PART II

A LANGUAGE FOR EVOLUTION
AND REVOLUTION

SEVEN
Q & A, AND A NEW LANGUAGE FOR EVOLUTION

At this point, roughly midway through the 20th century, we've reached what history now reveals was the pivotal juncture for the development of both scientific and social mind in the 20th century. For now, with the body blow impact of World War II, the atom bombing of Hiroshima and Nagasaki, and the Cold War threat of nuclear holocaust, the split in mind affecting more than science was moving into every aspect of the lives of our species and all species on this planet.

We're looking at what looms in retrospect as the crunch point for both the theory and story of evolution—the rough point in time where, with the addition of ever more rapid environmental devastation, the size of the threat to survival began to run so far ahead of theory that only a monumental blinding and hamstringing of both science and society could sustain the disparity.

Beyond the scientific board room the questions of the customers and shareholders in NeoDarwinism, Inc. are beginning to pile up. From here on rise the questions that science, now at times deeply concerned, began to nibble at during the final years for the 20th century, to which we'll turn in Part III. But far more important now, these are the questions that at last must be answered if the 21st century is to take us anywhere near where we *must* go if all embodied in the word humanity is to survive.

Why, for example—at this midpoint for the 20th century—had every attempt to free the mind and let it roam been fenced in, hamstrung, or otherwise grounded?

Certainly, one cannot fault the focus on prehuman evolution or on genes to provide the proper grounding for an understanding of human evolution. But by now it was obvious that something much larger is at issue.

As noted earlier, Darwin wrote 200 times of brain and mind in *Descent*. Romanes and Morgan had initially focused on the study of *intelligence* in animals. Why, then, was this thrust of interest so readily side-railed to focus almost exclusively on the study of prehuman, even hypothetically pre-sentient evolution?

Why was the concept that briefly surfaced as the idea of initiative by the organism, or "self-adaptation"—clearly articulated in Darwin over 100 years in advance of the development of self-organizing theory in our time—fiercely resisted, and only very late in the game accepted by a reluctant minority among Neo-Darwinians?

This omission is striking when we consider that in case after case the innovators themselves—Fisher versus Pearson, for example—were vivid examples of this factor of mind at work in their own drive to go up against the prevailing paradigm.

And why, if from Darwin on, among the scientists themselves the loaded term "survival of the fittest" was avoided or deplored, did it persist?

Why did this psychologically, socially, politically, economically, and morally destructive phrase become not only the main but, mostly and automatically, the *only* explanation for evolution passed along by the media and people in all walks of life?

Why did this fixation prevail despite the expenditure of millions of dollars and billions of woman and man hours on education globally?

And how and why was social science automatically and for so long excluded by the Neo-Darwinian monopoly for evolution theory? How and why can it be said that functionally the social sciences were "relegated, like children, to the sand box, while the grown ups went off to rule the world."[1]

And why did social science allow this? Why was opposition to the Neo-Darwinian monopoly almost exclusively grounded in biology so easily turned back and the voices of both social science and the humanities so readily stilled?

As I show in detail in *Darwin's Lost Theory,* along with Herbert Spencer, Darwin himself set out to provide the grounding for, and in his final years point the discourse toward, application of the vast range of science beyond physics and biology to cultural evolution and general evolution theory. Thereafter, as we've seen, from Romanes on came the spread of this embryonic Darwinian vision into development of the new reach of psychology, sociology, and the rest of social science on into a systems and evolutionary systems science. Thereafter, within social and systems science, erupted the questions, WW II and atom bomb driven, of customs, beliefs, ideology, values, and above all, morality.

Yet why, then, if there at the beginning was Darwin's architectural sketch for this larger reach *and* responsibility for the development of evolution theory, have we had to wait until the 21st century for the logical partnership to begin to form between natural and social science to move beyond the foundation to build the vital *superstructural* understanding of evolution?

And what is the place of religion in all of this?

At this point in our retrospective America had been captivated by the heady drama, in1925, of the Scopes "Monkey Trial," pitting Clarence Darrow and the pro-evolution "city folks" against William Jennings Bryan and the anti-evolution "yokels."

And what happened thereafter? With nearly a century of public and private education since then, why has Creationism grown until by 2009 a Gallup poll on Darwin's 200th birthday found that only 39 percent of

Americans believe in evolution versus 43 percent for God and Creationism?[2] Or that earlier polls found two-thirds of Americans want creationism taught along with evolution in schools?

All of which raises still another pertinent question. Why, with the exception of the years of Morgan's fruit flies, and the obligatory attention to Darwin and evolution every year on or around the time of his birthday, has time-wasting story of Creationists versus Evolutionists been about the only thing the media ever covers on evolution?

And why did so much of what was later rediscovered and acclaimed by his successors, remain buried in the body of the lost Darwin? Why had apparently only Romanes and Kropotkin notably perceived and declaimed the potential disastrous consequences of this loss?

When for Darwin his case for the moral sense as the prime driver for evolution at our level of emergence was a matter of passion, why during the years that locked in place the Neo-Darwinian monopoly was what constitutes our sense of right versus wrong mainly pursued by the tragic and scandal-ridden J.M.Baldwin?[3]

Finally, both obscuring and underlining all of this, is the question NeoDarwinism, Inc. forced the world to face: What is the connection between both scientific and religious theory and the raw impact of real world events. Increasingly bearing on where we are headed in evolution and the battle for human survival, it is the fundamentally moral question raised by the eugenics movement of *human engineering*.

What kind of human do "we" nurture or manufacture? What kind of society do we liberate or control?

This is the ultimate question running through all of the above. It's further the question we earlier raised of the over-riding something or other within the history we've examined, which at times seems everywhere to lurk around the corner. Beyond armchair philosophy or mere imagery, what is this dark presence that from time to time seems to act like a giant invisible puppet master dipping down or slipping in to nudge the discourse and the action invariably backward rather than forward in human evolution?

What could a new and bedrock practical level for evolution theory tell us of this force or process, which seemingly out of nowhere plops down on stage a Hitler, Stalin, Pol Pot, Idi Amin, or within democracies smiling puppets to front for the regressive Powers That Be?

In short, what drives us ahead, checks us in place, or drives us backward in evolution?

Could answers to these questions be found within the prehuman mindspace of biology prevailing for NeoDarwinian evolution theory?

Obviously not. And so fitfully but steadily the second Darwinian revolution gained momentum. The findings of progressive brain research, psychology, sociology, anthropology, economics, political science, *evolutionary* systems science—above all, a new science of moral mind and action, which out of the lost Darwin nudges at us today[4]—hammered at the gate to evolution theory demanding entry and room for expansion.

But to be heard you must speak with a language those you must reach can understand. In the rest of the chapters for this section we will look at the conceptual, semantic, and social challenge faced by the revolutionaries, but fuzzed by the counter revolutionaries. The problem of a language that across all fields could be understood.

EIGHT
OF BUDAPEST, FIELD, CHAOS, AND
COMPLEXITY THEORY

By late mid-20th century the conceptual explosion of cybernetics, chaos, complexity, and self-organizing theory—driven by the lock-firm evidence of the active brain, the excitement of the fluid new power of the computer, all embraced within the vital field of systems science—was underway.

I came to the wonderland of this new exploration of the concepts and languages of evolution via a real life episode out of a spy novel. The phone rang. It was a call from a strange voice with a heavy Hungarian accent. Could I come to Budapest?

The year was 1984, at the height of the Cold War, with Eastern and Western Blocs armed with sufficient nuclear overkill for the possibly final holocaust, on hair trigger alert.

There was to be a secret meeting behind the Iron Curtain in Budapest of scientists from both sides. Brain child of one of the most remarkable scientists of the 20th century, the Hungarian-born, ex-concert pianist and pioneering general evolution theorist Ervin Laszlo,[1] the idea was to see if global nuclear disaster could be averted by turning the power of chaos theory, then new and popular, to peaceful ends. The objective, it turned out, was to use chaos theory, and more generally systems science, to expand the theory of evolution beyond the old survival of the fittest thrust based on the deep past. Our goal was to bring together and integrate the new work of many fields embodying an emphasis on cooperation and common cause to serve nationally and globally as a scientific policy-shaping guide to a better future.[2]

Soon, after the ups and downs I will write of in the third book for this trilogy, *Up Against the Paradigm,* with Laszlo as founder and myself and others as co-founders, we launched the General Evolution Research Group, with members across the board from physics and biology throughout practically all the social sciences, from many nations, with a new scientific journal, *World Futures: The Journal of General Evolution.*[3]

As the Cold War years dissolved into the brief hope for a "Peace Dividend"—rather than what soon became all-out Greed for both sides, and preparation for more war—four things became evident.

First was how quickly the implacable weight of the status quo could sap the early surge of our bold notion of going up against and transforming the old paradigm.

Yet there remained the subtle power of the perspective of what among ourselves we called the Chaos Revolution. It was the power of this new mind and language that freed and allowed one to, in effect, stand aside from the old struggle. Coupled with the advance of systems science, it was the power to at last clearly see the battle within politics, economics—and increasingly within religion, or more generally spirituality—that was in fact light years beyond the gene or atom in shaping evolution.[4]

But here, too, was a problem to which I was especially sensitive. Having been a journalist and aspiring novelist, poet and playwright before shifting to science,[5] I could see that all that had been thrown at the mind space of science and the world at large was way too much to absorb and put to use. Autopoesis, dissipative structures, flow, Gaia hypothesis, holographic brain and mind, partnership systems, domination systems, quantum, zero-sum, Akashic field—out of the glut for this rich mind food we needed a simple and manageable basic handful of nutrients. Out of the

61

confusion of this new scientific smorgasbord, atop the glutted mountain of previous scientific concepts and languages, could one somehow identify a simple entry diet?

In my training as a psychologist I was fortunate to be a student of a research associate of the pioneering systems psychologist Kurt Lewin.[6] A refugee from Nazi Germany, major influence on the fields of social, child, organizational, and leadership psychology, founder of group dynamics, T groups, sensitivity training, action research, in his time Lewin's impact on psychology was considered the most important after Freud.[7]

All of this was loosely bound together in Lewin's writing, research, and teaching by his concept of *field theory*. Simply put, the idea is that just as everything within a baseball field in one way or another relates to baseball, or in a gold field relates to gold, in the same way our mind is segmented into fields within which we gather everything relating to what we want to accomplish and take care of.[8]

For example, to know and play baseball you don't just focus on pitcher, or catcher, or bases, or fielders. Your mind reaches out to grab all of this and everything else bearing on the game. In other words, your mind reaches out to gain a sense of the system, of the field as a whole.

Here is the disarmingly simple way the Lewinian approach begins.

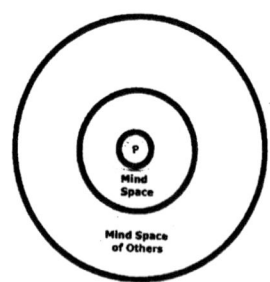

Figure 8.1
Lewinian Field and Mind Space

One draws a circle to represent the *person* (P). Then around it we draw a larger circle to indicate the spread out around a person for her or his consciousness, or more descriptively, *mind space.* Then we surround both person and personal mind space with a much larger space to enclose the consciousness or mind spaces and cumulative mind space of others.

The first time I saw Lewin's new visual language for psychology I thought it was the silliest thing I ever saw. It looked like nothing more than childish doodles on a page.

Yet from this kind of uncomplicated beginning Lewin developed what many became convinced was the most powerful and easily accessible language for understanding the dynamics of mind in action.[9]

The great advantage of this new approach to science was, and still is, that by shifting from the heavily entrenched verbal-based analytic power of the left brain to the visual-based systems power of the right brain, one could easily see, and quickly track, vital components and interactions for mind space at all levels, from the smallest to the largest in evolution.

For example, in Figure 8.2 can be seen how easily we can visualize the interaction of the mind space of the person; with the group; with the mind space of the World at Large.

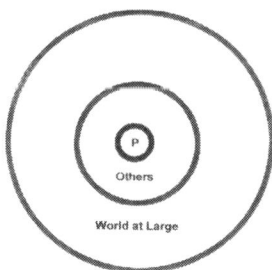

Figure 8.2

Relation of Person, to Group, to World Mind Space

What I think ... intersects with what you think ... intersects with what happens everywhere else on this cognitively shrinking planet to either drive us forward or backward in evolution.

In the past this was a tenuous and tedious process shaping the wobble for our species forward and backward over centuries. But in our electronically interlinked world of email, instant news, globally supersensitive to events, all this has exponentially changed.[10] The speed up for this process has become a force impacting the dynamics of evolution either favorably or unfavorably exceeding our power not only of social but also scientific coping.

In the 20th century the development of a new mind space and coping language for science initiated by Lewin in the 1940s was advanced by the mathematician Edward Lorenz in the 1960s. As the Cold War and nuclear overkill pitting the U.S. against Russia escalated toward global chaos, out of Lorenz' attempt to track the intricate movement of winds and clouds for weather patterns emerged what became modern chaos theory.[11]

Here, for the first time with reasonable ease, chaos theory made it possible to track the action of the tiniest of organisms and processes across the barriers of biology and physics into the largest of movements and patterns for the social sciences, the humanities, religion, and all else bearing on cultural evolution.

Here, further, was an immense advance in the power to perceive the pattern for movement from past, to present, to future, which constitutes everything we experience either as advance, progression—as, per se, *evolution*—or as being checked in place, or the regression we experience as lack of evolution.[12]

But most important—as Lorenz had demonstrated with the mathematics of weather patterns—chaos theory could provide a quick link to the crucial endgame challenge of real world impact.[13]

Here, for a quick sense of the advance in understanding evolution that

Lorenz offered, are illustrative patterns for chaos theory from the work of the mathematician and pioneering chaos theorist Ralph Abraham and his brother, the pioneering chaos psychologist Fred Abraham.[14] Here we may glimpse how processes involved in the simplest to the most complex movements take specific shape moving from past, into present, into the future.[15]

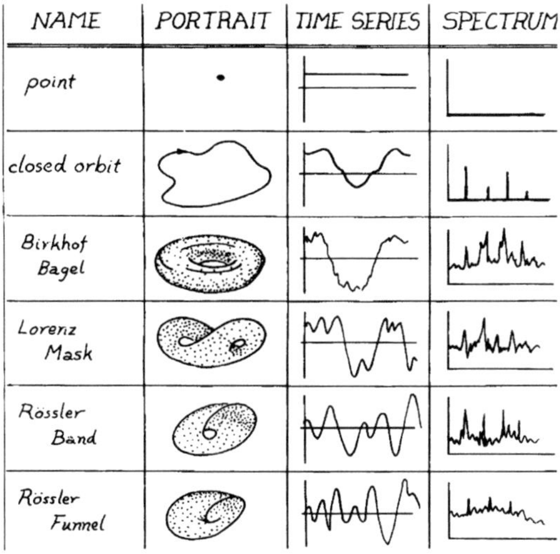

NAME	PORTRAIT	TIME SERIES	SPECTRUM
point			
closed orbit			
Birkhof Bagel			
Lorenz Mask			
Rossler Band			
Rössler Funnel			

Figure 8.2
Chaos theory configurations

One more concept—the Lorenz *attractor*. Again, the idea is basically simple. Just as the entry of an attractive woman or an attractive man into, say the waiting room for a doctor's office, creates a subtle force field pulling group attention toward them, so does the attractor operate in chaos theory.

A *static attractor* keeps everything fixed in place for a time, for example, the attractive one sits down. A *periodic attractor* alternates between two points of interest, e.g., two equally attractive men or women enter the room, with attention wandering back and forth from one to the other. Then there is the *strange or chaotic attractor*, which can arise suddenly out of nowhere to take over and pull everything within a field into whatever direction the strange attractor happens to move—as for example, what happens when one suddenly comes upon a movie star in a supermarket.

What first drew me to chaos theory was the recognition that here was a new science for both tracking and *predicting* evolution, or how we can advance the best and avoid the worst as we move into the future.[16]

The static attractor checks evolution in place. The periodic attractor is what century after century can be tracked in the alternation of liberal with conservative periods of history.[17] It is the strange or chaotic attractor, however, which provides the most crucial of thrusts. For here was a new language that could not only open the way to a new power for tracking whether we are being driven forward or backward in evolution. Here was new hope for prediction *and action*.

In other words, if through Lorenz and chaos theory we had gained a measurable and undeniable new power to better predict and better cope with the ravages of weather globally, why not also everything socially, politically, economically, morally and spiritually that drives us either forward, checks us in place, or drives us backward in evolution?[18]

Here, then, is a useful A, B, C for resuming our story of the ups and downs for 20th century theory and consequences.

The *field*.

The content within the field of the *mind space*.

And the vector or arrow point for forces driving mind and action backward, forward, or in other patterned directions: the *attractor*.

NINE
PUTTING THE NEW LANGUAGE TO WORK

At the time of the publication of Julian Huxley's *Evolution: The New Synthesis* in 1942, the single most devastating event of the 20th century was well underway. Europe was already three years and America its first year into World War II.

Thanks to effective military use of the destructive power of the sciences of physics, chemistry, and biology, by the war's end, in 1945, in Europe 60 million people had died, 20 million of them soldiers, 40 million civilians.

In the United Kingdom German bombing had devastated London, Liverpool, Birmingham, Manchester, Bristol, Belfast, Cardiff, Clydebank, Kingston upon Hull, and Coventry. In Germany British and U.S. Allied bombing had leveled two of the nation's most treasured cultural sites, Dresden and Nurnberg, and ravaged Berlin, Hamburg, and Cologne. In Asia, the Japanese attack on China left an estimated 20 million dead or wounded, with 95 million refugees. In Japan, the estimate was nearly 2 million dead, wounded, or missing. Then came the instant obliteration of 140,000 men, women, and children in Hiroshima, then 80,000 more in Nagasaki, as America's development of the incredible new nuclear power made it possible to level both cities with just two atomic bombs, "Little Boy" and "Fat Man."

Behind it all was the poisoned mind space, until then beyond imagination, fully revealed in the horror of the death camps, of Hitler's horrendous experiment in human engineering. It was only slowly uncovered afterward, but not only had six million Jews been snatched from their lives and gassed, or worked, or starved to death. Also signaling what was intended for the whole world, had the Nazi dream of global conquest succeeded, were the cooly meticulous death camp records of the murder of two million Poles and four million others deemed "unworthy of life"—the disabled and mentally ill, Soviet prisoners of war, Freemasons, Jehovah's Witnesses, gypsies, homosexuals.[1]

I write here not of just something I know out of books, after the fact. First, as a fledgling journalist in the U.S.Navy during the end of World War II, then later through my marriage to a holocaust survivor, I came to know more than is usual about what can happen if the wrong, or disastrously truncated, scientific *or* religious theory of human evolution infects the mind space and is put to use by the worst rather than the best among us.[2]

Not until the pictures of the death camps hit the world did it finally reach most people in America and elsewhere how evil the worst among us could be—or that evolution is not a linear process.

In other words, many of us still routinely assume the word evolution means slow, steady upward movement—that if we just keep plugging along doing the usual, things get steadily better over time. Now reality slammed across the picture of a *nonlinear* or curvilinear process.

We were forced to see the true picture of evolutionary reality: how with social hurricane force we can be driven backward, or dangerously checked in place. During the war years we also came to know something equally important: how with mind space inspired by progressive leadership, in science as well as in politics, economics, and religion, we can do better than just inch our way forward in evolution.

Can we map this process?

Can we uncover a scientific road map for human evolution?

Some possibilities fall in place if we shift from the mind space of biological evolution for a look at the movement of *fields* and *attractors* within the huge mind space of *cultural* evolution.

Here is a new look at the years between the great synthesis for the Neos and the rise of the Super Neos—and our choice of direction thereafter.

The Challenge of Cultural Evolution

Along with Kurt Lewin, out of the flight of Jewish intellect from what became Hitler's Europe came Albert Einstein, Sigmund Freud, T.W.Adorno, Erich Fromm, Else Frenkel-Brunswick, and other scholars of the remarkable Frankfurt School.[3] Also in flight was a seven year old Jewish child who 49 years later was to make a major contribution to our understanding of cultural evolution and provide a new language for it..

This was another co-founder of the General Evolution Research Group: macrohistorian and evolution theorist Riane Eisler..[4] Out of a multidisciplinary study of 35,000 years of human cultural evolution, Eisler developed the case for two models at work in shaping the cultural evolution of our species.

On one hand is the impact of a "partnership" model, systems, and ethos driven by the ideals of "power *with* others"—freedom, equality of gender and race, peace, sharing, caring, the valuing of linking more than ranking. On the other is the impact of a "domination" model, systems, and ethos driven by the ideals of "power *over* others"—top down control, inequality of gender and race, exploitation of others, violence as well as war on the grand scale to gain ends, the valuing of ranking more than linking[5].

Over thousands of years this polarity was discerned by deeply concerned religious visionaries—Jesus, St.Paul, Gandhi; by philosophers—Kant, Hume, Hutcheson; now within our time by scientists.[6]

The 20th century work of Adorno, Else Frenkel-Brunswick, and the

69

UC-Berkeley team, which contrasted the devastation of the *authoritarian personality* with the liberative thrust of the *democratic personality*, was and is pivotal.[7] Likewise was and is the work of anthropologist Ruth Benedict's contrast of the impact of *synergistic societies,* driven by the ideal of working together for the benefit of all, versus *non-synergistic societies,* driven by the ideal of me and mine above all.[8] Eisler, however—in a series of books revealing the contrasting impact of the domination versus partnership model on sex, politics, education, economics, personal and social action, and spiritual evolution[9]—both synthesized and went beyond prior work in uncovering the centrality of this polarity in the evolution and choice of future for our species.

In keeping with the visual thrust of a new language for evolution, here is my own perception of a core difference for the partnership versus domination mind space in terms of economic and political evolution.

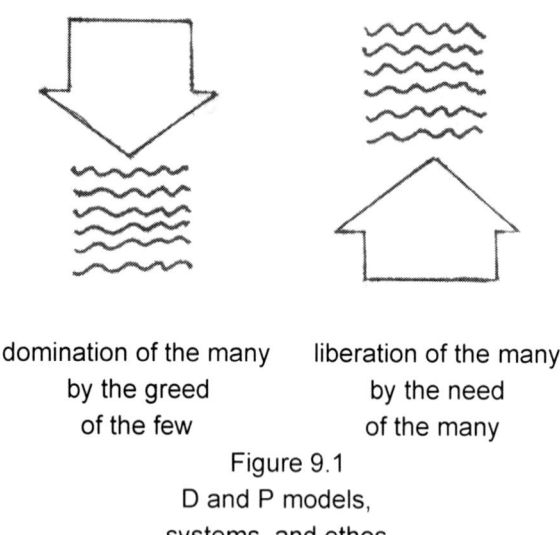

domination of the many   liberation of the many
by the greed   by the need
of the few   of the many

Figure 9.1
D and P models,
systems, and ethos

70

To put this polarity bluntly and directly within the context of languages for evolution, at the moral core we are looking at the *ethos of survival through sharing and caring* versus the *ethos of survival of the fittest*, i.e., *the most selfish and the meanist.*

We are looking at the conflict between these two mindsets and the battle over which mind space is to govern our choice of the future for ourselves and our planet.

TEN
EVOLUTIONARY POLITICAL SCIENCE

Swiftly now, by the 1930s—as shown by Hitler's self-justifying references to survival of the fittest Darwinism and Joseph Goebbels' genius for propaganda[1]—the evolution of our species was driven radically backward by the predatory manipulation of the mind space of Germany and the World at Large.

What among the Neo-Darwinians began as a scholarly discourse about eugenics explodes into the mass experiment of Hitler and the Nazis.[2] In other words, what looked like something no bigger than the proverbial cloud no bigger than a man's hand against the horizon in one mind space became a hurricane in another. On one small island of the global mind space were the Neo-Darwinians, secure in the amoral isolation of their truncated theory, speaking and writing of comfortably bodiless abstractions. Out in the mind space of the World at Large, however—that is, in the real world of politics, economics, mass ignorance, and regressive religion—were those looking for scientific justification for a reign of terror and ultimate uncontested power.

So through the change in meanings and consequences from mind space to mind space the disaster erupts. Natural selection and variation for the theorists, becomes survival of the fittest for the mind at large, becomes *Deutchland uber alles* and all its lesser versions for the power-mad pack of circling wolves. (See The Controversial Connection in ending Reflections for a vital semiotic analysis in visual language terms).

Now the flow of the relevant stream of mind moving from the past, through the present, into the future can be seen in the following configuration.

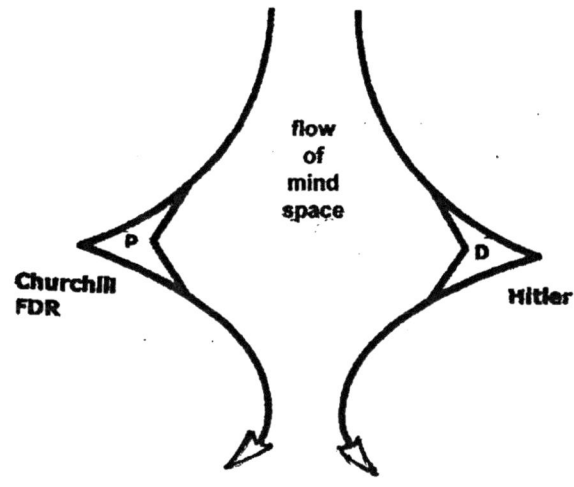

Figure 10.1
P-*attractor* versus D-*attractor* in
Political, Economic, Cultural, and Moral Evolution

However strange this configuration may look at first, visually it is just the round mind space of previous Lewinian figures stretched out to simulate the flow and spread of global mind space—as if we were mapping the semiotic shrinking and the swelling of a river moving through both space and time.[3]

In chapter nine we introduced the case developed by cultural evolution theorist Riane Eisler for two models at work in shaping the cultural evolution of our species. On one hand is the impact of a "partnership" model, systems, and ethos driven by the ideals of "power *with* others"; on the other is the impact of a "domination" model, systems, and ethos driven

73

by the ideals of "power *over* others."

In chapter eight we introduced the basic concept of the *attractor* from chaos theory.

In Figure 10.1 we now wed the two to simulate the thrust of partnership versus domination attractors, or *P-attractors* versus *D-attractors*.

Now in the swelling of the flow of global mind space into the years of World War II—advancing the threat in one direction, responding to it in the other—we see how the mind space for the world at large was in effect pulled out on either side by the armies, economies, allies, and all else symbolized by the names of Hitler in one direction, Winston Churchill and Franklin Roosevelt in the other.

Behind the visual rightward political thrust of the Hitlerian and fascist ethos more generally was the *D-attractor* mindset, driven by a vision of the "perfection" of our species through enslavement, suppression, sterilizing, or slaughtering all who fell short of or opposed the Nazi vision of the *ubermench* or superman.

Now throughout Europe, America, and Asia all that was struggling to become something more than political, economic, and religious lip service to *P-attractor* ideals suddenly becomes vividly meaningful.

To counter the expanding thrust of the Hitlerian *D-attractor* all that over thousands of years has mainly driven the evolution of our species became mobilized behind the expanding thrust of the *P-attractor*.

Behind the thrust of the Rooseveltian, Churchillian, and democratic ethos more generally was the partnership vision. Behind it was the vision of the advancement of our species, and life on earth, through the child-raising, family dynamics, schooling, and enlightened political and economic public policies of freedom conditioned by equality. Behind it was the drive of the always challenged but evolutionary inbuilt basic and cross-cultural morality of doing unto others as you would have them do unto you.

The result was what in chaos theory is called a *bifurcation*—or more concretely, as seen next—a swelling of mind to become a *split* into different directions, in this case one thrust driving us forward, the other bent on driving us backward in evolution.

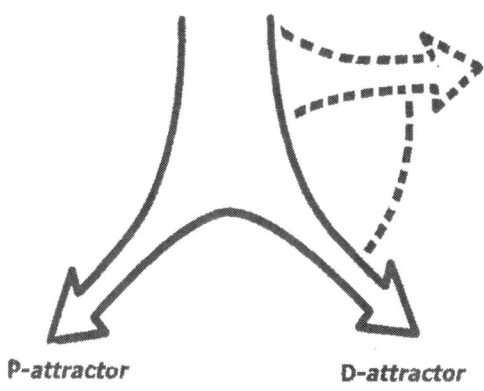

Figure 11.2
Mid-20th Century Bifurcation

In 1945 Germany, Italy, and Japan were defeated; Hitler and Mussolini killed; for a time much was made of the great evolutionary lesson embedded in World War II.

Driven by the P-attractor ethos, the United Nations and the World Court were formed. In 1947, accelerating the liberation of colonies, Gandhi and Nehru gained independence for India. In 1948, with formation of Israel, the Jews Hitler set out to eliminate finally gained a tiny patch of earth they could defend and call their own. But already implacably underway was the resurgence of the D-attractor thrust, with Stalin and the expansion of a new split for the global mind space into the nuclear-arming opponents of the Cold War.

Here, beyond evasion, the thinkers and leaders for our species were forced to realize we are not sweet peas or fruit flies. Nor are we cowed and dutiful humanoids with fate dependent either on the will of God or blind chance for Natural Selection and Random Variation. The threat of subjugation or death in the gas chamber or the gulags made it beyond question evident we are organisms with minds capable of a *choice* of positive or negative direction.

As W.E.Henley caught the ideal in *Invictus*, "I am the master of my fate: I am the captain of my soul."[4]

But here was the heart-rending tragedy for millions and a grim moral for the story. So firmly stuck within the Darwinian first half and the clutch of the D-attractor paradigm was most of the mind space for both natural and social science that the political, economic, and moral implications for evolution theory—and thereby national and global policy—bounced like rubber bullets off the armor of the prevailing mindset. And so for lack of the guidance that an up-dated, integrated, politically, economically, and morally and spiritually inclusive *progressive* theory and story of evolution could have provided, once again our history became the slaughter of more millions of our species and development of the atom and hydrogen bomb as "peace-makers."

Does this seem too extreme a charge, too big a leap in speculation?

Yes, of course, much else was involved. But let's take the next step and see what happens.

How closely does the configuration on the next page roughly approximate the political, economic, social, moral, and spiritual course of post-WWII real world cultural evolution up through the last years of the 20th century into the 21st century?

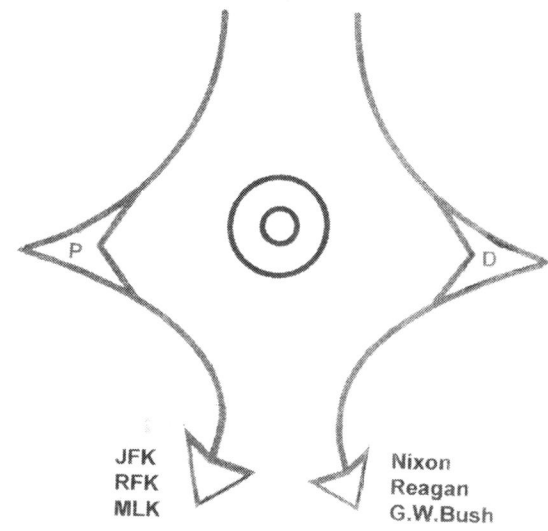

Figure 10.3
The Political, Economic, Social, Moral, and Spiritual Course
of Post-WWII Real World Evolution

Here, transcending the hot house world of an isolated evolution theory, can be glimpsed how science and society was and still is manipulated by D-attractor leaders and D-attractor followers to serve D-attractor ends.[5]

Now can be seen what lies behind what otherwise has seemed to be the dark magic of some giant invisible hand whereby, as in a puppet show, the engineering of the "proper" human and the "proper" shaping of human evolution is advanced.

In the stream of progressive mind space for America and the world at large symbolized, among others, by President Kennedy, Robert Kennedy, and Martin Luther King, and the stream of regressive mind space symbolized, among others, by Presidents Nixon, Reagan, and G.W. Bush, aren't we looking at the fundamental question of political alliance and evolutionary direction for the 21ˢᵗ century?[6]

Aren't we looking at something more meaningful than the customary multibillion dollar biennial football game with multibillion dollar beer and pretzels in the clash of liberals versus conservatives, progressives versus regressives, Democrats versus Republicans for the elections and policies that shape the evolution of our species and the future of our planet?

Aren't we looking at the politics of reason, caring, and sharing versus the politics of lunacy?

Surely by now it's evident we will continue to limp and stagger into the future as long as we lack the wider comprehension and warning of Darwin's completed theory and all the progressive *social* as well as natural science that both corroborates and advances it.

Isn't it evident we shall fall short, and invite oblivion, until we build the fully human, political, economic, and moral action-oriented theory and story now seeking form within the vision of Darwin's second revolution.

Despite many decades of philosophical and scientific dissent, science and society still remain primarily locked in the paradigm-preserving dictum that science must be inflexibly objective, neutral, above the fray.[7] But the bedrock systems dynamics of partnership systems and the P-attractor versus domination systems and the D-attractor underline the evolutionary requirement that if all that we value as humans is to survive, *we must choose sides, and for the better side, fight.*"[8]

This was and is the task to which science turned during the final decades of the 20ᵗʰ century.[9]

PART III

THE RISE AND FALL OF THE SUPER NEOS

ELEVEN
VISIONS OF THE SUPER NEOS

World War II has ended, Hitler's gone, the world now shudders under the shadow of the Cold War and the nuclear overkill of the United States and Russia. Beyond the Iron Curtain, however, has sprouted the drive of the 1960s for peace, tolerance, sex, love, and hope for the future the Beatles captured in music.

The mind space for America and the world at large is now the 1970s. As historians and political scientists unite in telling us today, the race-baiting Southern strategy has shifted power from the Democrats to the Republicans in America. Richard Nixon has been elected American president for a first term in 1968, re-elected in 1972, then in disgrace forced to resign in 1974.   The next year, within the mind space of science, onto the stage—or, if you prefer, into the board room for Neo-Darwinism, Inc.—strides Harvard eminence E.O.Wilson with *Sociobiology: The New Synthesis.*[1]

Swiftly this book became the bible for the new field of sociobiology. It also became the center of a firestorm of protest by sociologists, feminists, psychologists, and geneticists. Inflammatory passages in the book sparked outraged charges of sexism and racism.[2] Wilson's primary eminence was in the field of entomology, the study of insects—which further raised the question of his qualifications for pontificating on the human world.

Among most readers, however, such concerns were swiftly dispelled by the magnitude of his scholarship. In his passion for nature, the range of its subject creatures, and in his rare new emphasis on the urgent need for scientific attention to the question of morality and moral evolution the book and the new field spoke to rising concerns.

In keeping with the glaring omission ever since Kropotkin, again in this pivotal launching for the Super Neos there was nothing on *Descent* or Darwin on morality. But here at last, three quarters of the way through the 20th century, was the rare display of a firm philosophical grounding and far more than lip service to what for Darwin was the primary driver for human evolution. For Wilson provided an excellent summary of moral theorists and theories from the 18th century into the present.

Few so succinctly captured as much about the philosophical study of the "moral sense" as Wilson did starting on page 562 in *Sociobiology*. John Locke, Jean-Jacques Rousseau, Immanuel Kant, and the modern John Rawls he identified as "ethical intuitionists."[3]

As Wilson saw it, the problem was that for them the brain was a "black box," or one had to operate on guesswork as to its contents. To Wilson what was needed was a shift to "ethical behaviorism"—that is, strictly going by what a person does rather than what he or she supposedly thinks. Here his position—completely contrary to the lost Darwin—was that morality is learned as a form of behavior ultimately always shaped by selfishness, rather than in any way by any pre-existing inbuilt directive.

Opposing this position, Wilson noted, was the developmental-genetic conception of Lawrence Kohlberg and Jean Piaget, which I deal with at length in *The River and the Star*.[4] An important piece, however, was still missing. This missing piece Wilson felt that he and biologists Robert Trivers, William D. Hamilton, and zoologist Richard Dawkins were at last supplying with development of the new field of sociobiology.

Despite all the concern about the down side we'll come back to in Book II, one couldn't help welcoming in Wilson a mind capable of ranging beyond the amoral prison for lesser thinkers.

As few others had yet done at the time, Wilson brought consideration of moral brain structure into the picture. He linked morality to the "emotive centers of our hypothalamus-limbic systems."

This is true. But already by that time the research of Paul MacLean was showing this is true only of the lower or "pure selfishness" limbic level for the brain affecting morality. The higher, self-transcendent moral level of Darwin's greater interest—which, as I detail in *Darwin's Lost Theory*, the work of this great brain scientist confirms—emerges with the higher and later development of the frontal lobes of our brains.[5]

MacLean was delighted with and found striking the connection when I pointed it out to him. For the track that in the 1960s he uncovered in brain development from the emergence of the earliest reptiles up through mammals to ourselves, summarized here in chapter two, both mirrors and corroborates the theory for the stages of the development of the "moral sense" that Darwin first wrote of at the age of 28 in 1837, for later development in *The Descent of Man*.

MacLean's top brain placement for operation of the moral sense was also prefigured in Darwin. In another display of how Darwin's intuition ranged beyond the data of his time into what scientists late in the 20th century began to explore, Darwin remarked on the function of the brain behind "the frontal part of the skull—the seat of the intellectual faculties."[6]

So now, after decades of wandering without a forceful champion among the Neo-Darwinians, within the mind space of science, Wilson raised the vision of morality as a factor in human evolution to a high place among the stars.

Out in the political and economic mind space for America and the world at large, however, a new phase of the counter-revolution was taking hold.

David Loye

Trivers' Vampire Bats

In our "new language" chapters we developed the case for the use of field and chaos theory as source for a new language for evolution. We've seen how by using the concept of the *attractor*, particularly the domination or *D-attractor* versus the partnership or *P-attractor*, we can track how human evolution is advanced, or checked in place, or driven backward.

We've further glimpsed how, through the overlap of systems, of mind spaces linked in a chain one to another, the link is forged between what is said and done in the lofty or "clean" mind space of science and what is said and done in the "down and dirty" mind space of politics, economics, and religion.

Now in the everyday real world pressure of leadership, money, and faith, in the rightward, anti-New Deal gathering of political, economic, and religious forces, we are moving from the American presidency of Republican Richard Nixon toward election of Republican Ronald Reagan in 1980.

Concomitantly—as if only coincidental, but through systems interconnections in fact subtly linked—among Wilson's prized developers of the new field of sociobiology what soared with Wilson plunges like a dead weight back to earth.

If this were a movie it would be like a touch introduced by a Charlie Chaplin or a Harold Lloyd for comic relief. For moving on from sweet peas and fruit flies, the favored experimental animal for highly respected biologist Robert Trivers was the vampire bat. And so began twenty years during which the most highly publicized research for sociobiology set out to prove that human evolution was indeed, beyond all quibbling or question, driven by selfishness, core drive for the mindset of survival of the fittest.   From his careful study of the behavior of vampire bats and other creatures Trivers arrived at the widely hailed principle of "reciprocal altruism." This was the ostensibly scientific conclusion that everything we do for any one else is fundamentally motivated by what they can do for us.

In what became a cluster of related findings this message was hammered across by the work of William Hamilton, Richard Dawkins, and to both cap them off and set them straight, the French Resistance leader and Nobel prize winning biologist Jacques Monod.

Seldom noted then, but obvious now was this dark social product. For the research of the Super Neos that seized the media fed the heady story that science now had proof that altruism, or the desire to help others that liberals and progressives made so much of, shrank to little more than selfishness in the end.

This, in turn, became the automatic assumption of scientific support for the politics and economics of rule by the chosen few over the lesser many, or winners over losers— in effect an update for the old surefire motivator for the countless conquests of the past: to the victor go the spoils.

Moreover, while those who considered themselves up to date on things might revel in this new science that gave a moral face to greed, within religious America the equation of altruism with selfishness became further reason to fight Godless science and its vile theory of evolution tooth and nail.

Both slap in the face and thumb to the nose to over 250 million Christians in America,[7] by no stretch of the imagination was this what Jesus had said. And so was further fueled the wildfire spread of Creationism as a legitimate protest movement, with devastating political consequences.

Altruism Trivers defined as "An act that benefits another organism at a cost to the actor."[8] This was in contrast to selfishness, which he defined as "An act benefitting the actor at a cost to someone else."[9] Cost and benefit in both cases, he noted, was "defined in terms of reproductive success."[10]

While on the surface these definitions seem impressive, if we take a close look at what Trivers was saying—rather than, as is usual for the student or the faithful follower, letting them drift in and out of our heads with a nod—at least two aspects link to what was happening out in the mind space of America and its impact on the world at large.

First, it's important to notice how Trivers' definitions both set the conditions for and guaranteed findings that would substantiate only the "lower" half of the supposed motivation for altruism.

On checking what Darwin had actually said, I 'd found the lost Darwin differentiates between two levels or kinds of motivations for what we call altruism today. One is selfishness, which he called the "base principle" accounting for "the low morality of savages."[11] This he found widely operating among us, no doubt about it, affirming that it *does* motivate us to help others, as Trivers claimed.

The other level for Darwin, however, was the higher principle of the "moral sense" embedded within us biologically over millions of years. Core to the "lost" completion for his theory, this is the thrust that not only impels us to again and again go beyond selfishness to help others, but is also the prime driver of evolution at the level of human emergence—which Trivers *by definition excludes.*

This is an old game for scientists. Define what you want to find in a seemingly impressive way, and then, lo and behold, you have found exactly what you said was the case!

Hamilton's Honey Bees

W.D. Hamilton was considered one of the most important of

evolution theorists at the time of his death in 2000. Among his fellow biologists and within the new field of sociobiology he was most highly respected for solving a problem Darwin had identified.[12]

Briefly stated, the problem was this: If natural selection favors the survival of the fittest, why do so many organisms including ourselves sacrifice themselves for the good of the group or for others? Going strictly by Neo-Darwinian theory, evolution should uniformly produce selfish, not altruistic, behavior across the levels for species.

Darwin had raised the question in regard to the close cooperation one could observe in colonies of bees. As the best alternative to Trivers' principle of "reciprocal altruism," earlier Hamilton had arrived at the more restricted principle of "kin selection altruism."[13]

On closely studying the mating of bees with their Queen, Hamilton found a variety of strategies were used by the bees to favor the production of kinfolk rather than strangers. In other words, underlying what on the surface appeared to be cooperative or altruistic behavior, he reported a subtle battle to restrict the gene pool strictly to offspring to which one was related. Queens, for example, ate the eggs of those to be excluded. Most impressive to all those who could understand his advanced mathematics was Hamilton's rigorous demonstration of the power of numbers to certify how his observations of evolution in the bee world were Writ Large for us in the human world.

Thus, a mother will protect her child before her sibling's child, her sibling's child before a cousin's child, and a cousin's child before a stranger. Hamilton's formula $C < R \times B$ became next step up for the mathematics of evolution.

That is, **C**ost in fitness to the actor is less than ($<$) the genetic **R**elationship between the actor and the recipient times ($\times$) the fitness **B**enefit to the recipient.

87

Again we have a finding that out beyond the tidy world of science could be welcomed and put to work toward political and economic ends. Here was a science that made it obvious the inheritance tax was an abomination that should be abolished. In fact, if this were a sensible world, all taxes, which unfairly cut back the capacity of the rich to propagate the better sort, should be abolished. Voting should be restricted, et cetera, et cetera.

Above all, during what became known among historians as the Greed Era, this weighty combination of Wilson plus Trivers and Hamilton seemed to further certify that selfishness is good. For wasn't this impressive support for the "trickle down" economics of the Reagan years? Wasn't this scientific proof that selfishness helps others in the end?

While lacking the huge investment in mathematics, which not only lay beyond Darwin's time, but for which he would have had no aptitude, once again in the lost Darwin I found Hamilton's explanation for altruism confirmed.[14]

But here again is the critical difference to which the Super Neos were emotionally, cognitively, socially, and scientifically blind.

Darwin specifically divided his analysis of altruism (i.e., mutual aid) into a *lower* selfishness and a *higher* "do unto others" level—and then went further only to be further ignored. He labored to counter the lower with the over-riding observation, and over-riding case made, for the transcendence of selfishness through the operation of "higher agencies." In other words, Darwin was saying the drive from lower to higher is an obvious matter of a growing up process for ourselves, *and for our species.*

Increasingly the main lesson of one billion years of evolution seems to be we mature or perish.

Dawkin's Selfish Genes

Trivers and Hamilton mainly produced articles for journals that few outside their field read. Their tremendous influence came about through what others wrote about them and the wonders of selfishness for doing good in the world. Next to the plate, however, came a zoologist with an uncanny gift for reaching beyond the sacred inner core to make science come alive to a wide readership.

In 1976 Richard Dawkins made his mark on history with publication of *The Selfish Gene.*[15] By the year 2000 *The Selfish Gene* had sold over one million copies, with translation into 25 languages. As further tribute to his impact, as a takeoff on the long established image of T.H.Huxley as Darwin's bulldog, Dawkins was dubbed Darwin's Rottweiler.

In retrospect, it's striking to see how many of the reasons for the backward thrust for sociobiology are revealed in the saga of Dawkins and *The Selfish Gene.*

To begin with, here was the triumph of an idea that could never have been taken seriously had American education been up to European standards. For it was like Hans Christian Anderson's story of the emperor served by the rogue tailors, who weave for him invisible clothes ostensibly of gold, who then parades naked through the admiring throng until at last a child cries out "but he has no clothes!"

It could be said that here was an uncomfortable kinship to an idea not only ridiculed in the early, quasi-scientific years of evolution theory, but picked up three quarters of the way into the 20th century and lionized: the so-called "selfish gene," successfully hailed as the answer to everything.

The earlier idea was that evolution was transmitted by a tiny replica of the human being within the sperm of males called the homunculus. Now it seemed that for Dawkins the gene was to serve the same function as the transmitter for not just any kind of human being. Most specifically, it seemed to be tailored to order for what politically and economically was getting underway for reliable production of the *selfish* human being.

Once again, in the cloistered mind space of science the idea wasn't this foolish or simplistic. But if we bring more new language for evolution into play, it can be seen how and why out in the woefully confused and adroitly manipulated mind space of America and the world at large this is precisely what Dawkins' title and presentation was widely taken to mean.

In other words, within the field for scientific mind the idea of a so-called selfish gene could be approached with all the intricacies, qualifications and rigor with which scientists advance their discourse. But out in the wide open space for the field of popular public and leadership mind the image of a greedy selfish gene with the power to govern all of evolution not only could take wide hold. It could swell out into the mind space of America and the world at large until it had become a significant factor in the thrust of the *D-attractor* driving the mind of America and the world at large in the regressive political and economic direction increasingly taking hold..

Should this be hard to picture, two more terms for a new language for evolution theory may clarify the situation. From the cybernetic theory of mathematician Norbert Weiner and others, widely used in management and systems science, comes the operation of *feedback* and the *feedback loop*.[16] From the sociology of Swedish economist Gunnar Myrdal and others, the other term is the operation of *vicious cycles* versus *virtuous cycles*.[17]

A book is written, for example, and takes hold. Back to the author comes the *feedback* of purchase, acclaim, high lecture fees, etc. Whatever thereafter happens is that, if one writes, or is, what the prevailing political, economic, and social system is looking for, there is set in motion a two way, mutually-reinforcing *feedback loop* embracing author and social and economic policy makers. Depending then on the direction toward which the loop is pointed, it can become either a morally backward-driving vicious cycle or a morally forward-driving virtuous cycle.

On one level, for example, success for Dawkins' books was a tribute to a rare ability to capture the essence of a scientific idea with striking imagery, and to write with a clear, strong, and engaging style. On a subliminally more subtle and powerful level, however, Dawkins' work became a prime case of the use of science to help justify regressive social and economic policy during the two term Ronald Reagan presidency in America.

Dawkins went on beyond implanting the ideologically-loaded idea of a "selfish gene," to success for implanting *The Blind Watchmaker* as an image for the Neo-Darwinian theory for the operation of Natural Selection and Random Variation, thereby further subtly reinforcing the "survival of the fittest" mindset.[18] Also, in tribute to his gift for influencing a wide readership, came to be his use of the "meme" as an analogue for the operation of the selfish gene and survival of the fittest at work everywhere within our minds and the mind of the world at large.

There was, however, no joy in one chunk of his readership. Among political scientists and sociologists was mounting concern about how sociobiology was being used to advance the political power of all that, within the perspective of a new language, was cumulating within the wake of the thrust of the regressive D-attractor. And in at least two other regards biologists found Dawkins' thinking uncomfortably slippery. There was discomfort (along with envy!) with the way Dawkins could simultaneously play the two fields—that is, the hypothetically pristine field of scientific mind and the inevitable distortion of scientific concepts by ideology out in the field of popular mind. Further, it was frustrating that he could make a lot of money doing this while claiming he wasn't saying what he was saying.

Another reason only now becomes apparent. It is that the work of Dawkins and the unreconstructed early sociobiologists not only served to further invisibilize what Darwin really said. With the most profound of consequences, it was implanting precisely what Darwin himself was foursquare *against* as the holy gospel for evolutionary science.

The subtle transvaluation from the mind space of science to the mind space of politics, economics, and the raging of regressive religion, was sharpening the drive within the mind space of America and the world at large toward the culminating global economic disaster of the G.W.Bush years. It was, in short, a process of *systematic transvaluation* accelerating the battle for 21st century mind and long term survival.[19]

What was Dawkins saying in *The Selfish Gene?*

We must teach our children altruism because "we cannot expect it to be part of their biological nature," he tells us.[20] In a world generated by successful genes driven by a "ruthless selfishness," "much as we might wish to believe otherwise, universal love and the welfare of the species as a whole are concepts that simply do not make evolutionary sense."[21]

Obviously no one who had ever bothered to read Darwin on the subject could have written this. Having unloaded the substance of his thoughts about morality in a few stray lines in *The Selfish Gene*, in *The Blind Watchmaker* in 1987 and *River Out of Eden* in 1995, Dawkins avoided tackling anything further to do with altruism, morality, ethics, or values.

It just seemed to be a matter in which he was wholly uninterested.

The Heights, the Depths, and the Ultimate Challenge

So now during the years of the movement from the years of the Nixon, to the Reagan, to the G.W. Bush American presidency, we have this new science that with a noble, sincere, and notably heart-felt vision of service to humanity E.O.Wilson set in motion within the mind space of science. But we also have the blind plunge thereafter into the mutually reinforcing dynamics of the vicious cycle swelling the backward thrust for the regressive D-attractor within the mind space of America and the world at large..

If it should be thought this is an exaggeration, one quote from all the books and studies that flowed together to become the *oeuvre* of the Super Neos underlines the message.

In retrospect, it seems to foreshadow the attitude toward the "do-gooder" that lay behind the smirk that became the badge of office for some of the prime agents for the D-attractor, e.g., American Vice President Cheney, Defense Secretary Rumsfeld of "stuff happens,"and most characteristically the president himself during the G.W.Bush presidential years.

The quote is from *The Economy of Nature and the Evolution of Sex* by Darwinian eminence Michael Ghiselin. Of the moral sensitivity of the "altruist" that Darwin celebrated as the central driver for ourselves and for human evolution, Ghiselin tells us this:

"Given a full chance to act in his own interests, nothing but expediency will restrain him from brutalizing, from maiming, from murdering—his brother, his mate, his parent, or his child. Scratch an 'altruist' and watch a 'hypocrite' bleed. No hint of genuine charity ameliorates our vision of society, once sentimentalism has been laid aside. What passes for cooperation turns out to be a mixture of opportunism and exploitation." [22]

Three years earlier, in 1971, the book *Chance and Necessity* by Nobel prize winning French biologist Jacques Monod put such a mindset to shame.

> Where then shall we find the source of truth and the moral inspiration for a really scientific socialist humanism, if not in the sources of science itself, in the ethic upon which knowledge is founded, and which by free choice makes knowledge the supreme value—the measure and warrant for all other values? An ethic which bases moral responsibility upon the very freedom of that axiomatic choice. . .[23]

But then to this soaring statement of faith in science Monod added the bleak conclusion, which asks our battered and bewildered species to welcome being left out in the cold with nothing but the science of the Neos and the Super Neos for comfort, alone in the universe.

> The ancient covenant is in pieces; man knows at last that he is alone in the universe's unfeeling immensity, out of which he emerged only by chance. His destiny is nowhere spelled out, nor is his duty. The kingdom above or the darkness below; it is for him to choose.[24]

TWELVE
THE BATTLE FOR MIND OVER MATTER

Earlier we met Julian Huxley and Theodosius Dobzhansky, co-chairs in creation of the Neo-Darwinian synthesis, which locked in place the mind space for the counter-revolution. Now we're to see the startling switch to revolutionaries.

The revolution was Darwin's original vision. Beyond the well-known first half, it was the lost moral action-oriented completing half for his theory of evolution. It was the vision for Darwin of the liberation of responsible mind Romanes died trying to uphold. It was the lost grounding for thousands of modern studies fighting for the same end.

The counter-revolution was what happened when first the Neos and then the Super Neos became fixated on the first half, and locked in by the thrust of the dominator attractor and the sweep of the domination paradigm could go no further.

This makes the shift of sides by Huxley and Dobzhansky so striking. Driven by minds of a scope that refused to be constrained within the Neo-Darwinism they helped lock in place, they chafed and muttered for a while. Then tired of trying to reform the prevailing paradigm, they decided to go up against it.

Huxley's Heresy

In 1946 Julian Huxley moved out of the world of sweet peas, fruit flies, and vampire bats into the bloody macrocosm for our troubled human world to become the first Director-General of the United Nations

Educational, Scientific, and Cultural Organization (UNESCO). Soon thereafter he began to spread the discomfort that made him no longer a good choice for invitation to parties in the tidy world of biology. Although the connection to Darwin was wholly unknown to him then as well as throughout the rest of his life, it was as if he'd suddenly been possessed by Darwin's ghost. He began to assert what actually had been the fundamental structure and key points for the long-ignored completing half of the founder's full theory.

The opening jolt for Huxley's heresy was his insistence on the need to move beyond natural selection to an exploration of *psychosocial selection.*

To some of his old associates that strange new word "psychosocial" must have sounded like something to do with the nut house, hence at all costs to be avoided. But Huxley clarified. Yes, natural selection exists and massively operates, no doubt about this whatsoever, he assured everybody. But the point was that at our level of human emergence *psychosocial selection* takes over.

"Though natural selection is an ordering principle, it operates blindly," Huxley observed. It "pushes life onwards from behind." It "brings about improvement automatically, without conscious purpose or any awareness of an aim."[1]

Psychosocial selection also "acts as an ordering principle. But it pulls [us] onwards from in front. For it always involves some awareness of an aim, some element of true purpose." In psychosocial evolution "the selective mechanism itself evolves as well as its products. It is a goal-selecting mechanism, and the goals that it selects will change with the picture of the world and of human nature provided by [our] increasing knowledge."[2]

Within the slippery world of ideas over which evolutionists were battling, Huxley set forth a clearcut vision of three distinctively different parts to evolution theory —cosmic, biological, and human. Or in terms of the scientific disciplines involved, physics, biology, and the social sciences.

Most striking in retrospect is how Huxley honed in on the primary concern that was both Darwin's and E.O.Wilson's in initiating the field of sociobiology. In 1943 his humanistic vision of the centrality of moral sensitivity in human evolution was eloquently set forth in *Evolutionary Ethics*,[3] in 1947 in *Touchstone for Ethics 1893-1943*[4], and in 1964 in *Essays of a Humanist.*[5]   In these sources we encounter what surely by now must seem the obvious fact I uncovered in the lost Darwin: a basic structure for any adequate theory of evolution consisting of two distinct parts. That is, Huxley separates the evolution of living systems into a first half applying mainly to prehuman evolution and a second and completing half applying mainly to human evolution.

This was a departure from the security of the world of test tubes, statistics, small animals, and the distant past that was hard enough by itself for his original cohorts to handle, but even worse was yet to come.

Dobzhansky and Teilhard de Chardin

Like Huxley, Dobzhansky also had misgivings about the exclusively biological embedding of Neo-Darwinism.

Particularly haunting is the following paragraph in which—again with the prevailing lack of knowledge of any connection to the rest of Darwin—in 1968 Dobzhansky expressed the perspective and even prime factors for the lost Darwin 100 years earlier.

"The most significant product, and the paramount determining factor, of human evolution is culture. The relationships between the biological evolution and culture are frequently misunderstood, and it is important to make them clear. Culture is not transmitted biologically through some special genes; it is acquired anew in every generation by learning and instruction, in large part through the medium of the symbolic language. "[6]

If right from the beginning the social sciences had been granted as firm a place in evolution theory as Huxley and Dobzhansky now insisted might the course of history have been different? Could we, for example, have avoided both Hitler and America's shameful Bushist entry into the 21st century?

A first half for Darwin's theory outlined in the realistically brutal biology of *Origin of Species*, a second and completing half outlined in Darwin's inevitable, logical, but "lost" attempt to move on to psychological and cultural evolution in *The Descent of Man*: it seems obvious now. Yet as we've glimpsed, the recognition of all there was to be gained by a good working partnership between natural and social science in the study of evolution was not only not favored but actively diverted, undermined, or outright killed off by social hurricanes in the wake of the D-attractor.[7]

Social science has never been popular in extreme or even mildly regressive regimes. Historically the social sciences and other fields bearing on cultural evolution arose by chopping away at the tyranny of the Emperor, the King, the Church, on at the Corporatism of our time; and indeed, from high to low, the bloody dynamics of all other domination systems.

Social science was, and is, and I pray may always be, the enemy of regression. Biology and physics, by contrast, from the beginning have been more easily manipulated to do the prevailing system's bidding.

Thus we have the tragedy of how Neo-Darwinism became a systems-crafted monopoly for biology. It was as if biology was shaped to serve as a Trojan horse loaded with the shock troops of "survival of the fittest" in its belly. Or in terms of a new language for evolution, as if it was co-opted to serve as an arrow point for the trajectory of the D-attractor, with the spread of domination systems in its wake.

But with Dobzhansky a glimpse into something even worse for the overlord was in store. Over thousands of years the sacred rule for domination systems has been Divide and Conquer. Hence, a vital systems strategy has been to encourage the battle between science and religion in all possible ways.

In this regard, Dobzhansky occupied much the same position as a gay among straights in the days when remaining in the closet, or at least maintaining a low profile, was the tactic for survival. For among a congregation of agnostics and atheists, Dobzhansky was a devout Russian Orthodox Christian.

Unless greatly compelled otherwise, you certainly didn't want to do anything to call attention to yourself in this regard.

Enter stage left, Teilhard de Chardin. A fascinating combination of Catholic priest and anthropologist, de Chardin had developed a theory of evolution combining science with spiritual and moral development. For this dreadful heresy he was not only proclaimed beyond the pale by science but forbidden by the Catholic hierarchy to publish his writings throughout his lifetime.

And Dobzhansky, to pile heresy atop heresy, had become a leading member of the American Teilhard de Chardin Association. In a 1973 article lambasting Creationism, "Nothing Makes Sense in Biology Except in the Light of Evolution," he quoted de Chardin to demonstrate why the barrier between science and religion on evolution was senseless.

Evolution, de Chardin had written, "is a general postulate to which all theories, all hypotheses, all systems must henceforward bow and which they must satisfy in order to be thinkable and true. Evolution is a light which illuminates all facts, a trajectory which all lines of thought must follow—this is what evolution is."[8]

But in regard to de Chardin, Huxley's heresy was even worse than Dozhansky's. With really astonishing boldness, when you think of the kind of courage this required within the context of his position and friendships in biology at the time, Julian Huxley was so taken with de Chardin that he wrote the foreword to de Chardin's main book *The Phenomenon of Man*, posthumously published in 1959.

Like Dobzhansky, Huxley veered toward the apostasy of Teilhard de Chardin because of the way it seemed to open the "higher path" that both he and Dobzhansky felt was missing from the Neo-Darwinism they helped create.

Soon, however, as eminent biologist Stanley Salthe notes in *Development and Evolution*, Huxley and Dobzhansky were not only *personae non grata* to parties for the old gang. They were also the sort one began to drop from one's references.[9]

Stephen Jay Gould, Darwin's Whipper Snapper

By 1980, ranked against the rise of the Super Neos, were two bands of adversaries. One was composed of the visigoths of Creationism. Contending that anything Darwinian was either evil incarnate or ungodly balderdash, their tirade was the beloved fodder of the news hounds. The other band was composed of the crusading anthropologist Ashley Montagu—whose warm, wide-ranging wit and scholarship we'll revisit in Book II—and paleontologist Stephen Jay Gould, genetic biologist Richard Lewontin, sociologist Steven Rose, and psychologist Leon Kamin. Their contention was that both sociobiology and its offspring, evolutionary psychology, were scientifically naive and potentially dangerous misapplications of biology to higher levels of human emergence.

The books of Stephen Jay Gould came to be valued by many readers as among the most delightful and erudite of the time. Gould became a major factor in bringing Darwin and his works to life. Along with Ashley Montagu, Gould was not only sensitive to the political dangers, but rare for the traditional scholar bold in speaking out and writing about them[10].

Montagu with *The Nature of Human Agression* in 1976, [11] and Gould with *Ever Since Darwin* in 1980[12], noted how beginning with late 19th century Social Darwinism emerged an eruption of books about killer apes, naked apes, books by Carleton Coon, Arthur Jensen and William Shockley about IQ differences, either subtly or openly promoting imperialism, sexism, and racism under the guise of science.

What they had in common, Gould cogently observed, was the "crude biological determinism" beginning to cloud the rise of sociobiology at the time of this first critique.

In the midst of the customary courtesies and niceties of academia—which so effectively preserve the status quo—Gould was refreshingly blunt in exposing the personal, social, and systems-functional motivations for purportedly Darwinian books of this type.

"They range, I believe, from pedestrian pursuits of high royalties for best sellers to pernicious attempts to reintroduce racism as a respectable science." Biological determinism, he observed, "has always been used to defend existing social arrangements as biologically inevitable ...Why else would a set of ideas so devoid of factual support gain such a consistently good press from established media ..."[13]

Sociobiology purportedly was setting out to promise more this time. It was being impressively expressed by scientists of the stature of E.O.Wilson. But if we turn to the expanding new perspective on evolution now possible, we can see what Gould, and even more so Montagu, with his early allegiance to Kropotkin and identity as a Jew during Hitler's years, were picking up.

Like the legendary canaries in a coal mine, they were impelled to serve as an early warning system to what could be headed our way with a new spread of the D-attractor throughout the late 20th century field of mind.

Both warned that one must keep a wary eye on what happened with sociobiology. Gould hoped that "the pluralistic spirit of Darwin's own work will permeate more areas of evolutionary thought, where rigid dogmas still reign as a consequence of unquestioned preference, old habits, or social prejudice."[14]

But what thereafter happened was a replay of the down side into the years that now are ours.

Lewontin, Rose, Kamin, and Free Will

Following Gould's first punch at the sociobiologists in 1980, biologist Richard Lewontin, sociologist Steven Rose, and psychologist Leon Kamin followed in 1984 with what many hoped might prove to be a knockout blow in the classic one-two punch technique for boxing.

The intended right to the jaw was their book *Not in Our Genes*. Insufficiently recognized then, but obvious now, were chapters of crucial insight exposing the social and political consequences of unreconstructed sociobiology. In the closing chapter, in a comparison of the new biology and the old biology and the relation of ourselves and all other organisms to our environment, the authors focused on the task of joining the first half to what I uncovered of the lost second half for Darwinian theory.

"Organisms do not simply adapt to previously existing, autonomous environments," they observed in a manifesto for the view of ourselves as active agents in shaping our destiny. Expressing what Darwin earlier showed in the invisibilized completion of his theory, we "create, destroy, modify, and internally transform aspects of the external world by [our] own life activities to make this environment."[15]

Lewontin, Rose, and Kamin further showed how the concept of free will—which ever since Immanuel Kant has been central to the theory of morality—is not the illusion sociobiologists and biological determinists in general tended to claim.

"Our brains, hands, and tongues have made us independent of many single major features of the external world," they observed of the human difference. "Our biology has made us into creatures who are constantly re-creating our own psychic and material environments, and whose individual lives are the outcomes of an extraordinary multiplicity of intersecting causal pathways. Thus, it is our biology that makes us free."[16]

Howard Gruber and Charles Darwin, Psychologist

The first book that began to crack the marble of Darwin's tomb was Howard Gruber and Paul Barrett's *Darwin on Man*.[17] Through psychologist Gruber's painstaking labor of many years, for the first time Darwin's stature as a social scientist became evident. Darwin the psychologist emerged and one could glimpse his fledgling sophistication as a systems scientist.

Particularly eye-opening was the attachment to their book of the full text of Darwin's early notebooks. Unpublished for 132 years, this was the first time this startling new material became available to a reasonably wide readership. Jotted down by 28-year-old Darwin at the early high point for his creativity, in the white heat of thought just after he returned from the voyage of the Beagle, to most readers this has seemed no more than a jumble with a spark here and there. However, it was in Darwin's notes on moral evolution I found the key to unlock the door into the full Darwinian realm—both the ground-in-to- the-point-of-banality first half to his theory, and the invisible completing second half.

Gruber and Barrett's book did not move on to link the moral theory in the notebooks with Darwin's carefully expanded and extensive statement in *Descent*. Nor did it so much as hint that in this connection we are looking at the completion for his theory of evolution. But in retrospect can be seen a vital contribution to recovery of the rest of Darwin and liberation of the mind of our species.

If here, three quarters of the way through the 20th century, the rest and possibly the best of Darwin could still remain buried, the incredible power of the overriding paradigm still imprisoning both science and society was starkly revealed. In varying ways Huxley, Dobzhansky, Montagu, Gould, Lewontin, Rose, Kamin, and Gruber and Barrett often mirrored what the lost Darwin was saying earlier, yet in none of these battlers for mind over matter did I find evidence that any of them knew it.

That is, in none of these champions for what Darwin had earlier observed and believed did I find evidence that any of them ever read—or if they had, that it had registered—the long bypassed sections of *Descent* I pulled together and first republished in *Darwin's Lost Theory*.

If other words, if such a hole in the mind space of this group of exceptionally wide read and knowledgeable scholars could exist, is it any wonder we could veer so far off track in evolution.

Robert J. Richards and Darwin,
Major Moral Theorist

Given the avalanche of books on Darwin during the 20th century, it seemed that surely scholars had probed and discovered and rediscovered everything of any importance whatsoever in Darwin—including, one is tempted to say, what kind of toothpaste or shoe polish he might have used. Yet, as I detail in Book II, in all of that century I could find only twelve books that referred to Darwin's moral theory, only four of which revealed anything beyond the most superficial and peripheral recognition of it.[18]

Other than Kropotkin's mainly out-of-print *Ethics*, in all of that century so badly in need of what Darwin labored so hard to give our species, Robert J. Richards' *Darwin and the Emergence of Evolutionary Theories of Mind and Behavior*,[19] first published in 1987, was the only book I could find that went into Darwin's moral theory in any depth, with fervent, real and respectful interest.

Richards was and is a psychologist and historian of science at the University of Chicago. His book was a masterwork of research exceeding even Gruber and Barrett's labors. It remains the indispensable source for understanding Darwin's moral theory in relation to its times and to the evolution of evolution theory.

"Darwin expended considerable effort on a theory of moral evolution, because he judged the moral sense, or conscience, to be by far the most important distinguishing feature of human nature," Richards wrote.[20]

"Darwin's method of approach had already been established during the period of his great creative effort, from late summer of 1838 through spring of the next year ... He now resurrected those early ideas, but altered, reformulated, and greatly refined them."[21]

We learn of the impact on the teen-age Darwin of his long walks and talks with the Scotch moral philosopher Sir James Mackintosh. Widely overlooked, it can be seen these walks and talks with Mackintosh were of pivotal importance. For Mackintosh was the leading heir to the great heritage of the Scotch Enlightenment school of moral philosophers including David Hume, Francis Hutcheson, and Adam Smith.

105

This was the same Adam Smith hailed as the father of capitalism today, whose *Theory of Moral Sentiments* and warning against the dangers of an amoral capitalism were buried almost as effectively as the rest of Darwin. It was via those walks and talks with Mackintosh that the open, questing mind of the young Darwin became grounded in the philosophy and theology of the moral sense for which he was to provide the scientific grounding.

We further learn of the inspiration of the great moral philosopher Immanuel Kant in the formation of Darwin's concept of moral mind. In Richard's book, we follow the exploration of the moral sense by Darwin's friend and rival Herbert Spencer, who coined the disastrous phrase Darwin regretted ever using, "survival of the fittest." We see the passing of the baton from Darwin to Romanes, from the dying Romanes to his student Lloyd Morgan, and from Morgan, in crossing the Atlantic to America, to Henry Osborn, William James, and J.M.Baldwin in the first attempt to break out of the prison of a reductionist biology with a mindful measure and theory of evolution.

It is a magnificent story magnificently told. But even in Richards we find the ambiguity that rises from the difficulty of trying, during a time of transitional struggle, to see the Darwin locked in place with the polished lens of the old paradigm through the still clouded lens of a new and better paradigm.

I will return to this problem in Book II. For an important aspect of the story of the exploration of evolution in the 20th century is the surprising contradictions, or switch in sides, among those who mainly filled the role of either revolutionary or counter-revolutionary during this surrealistic crunch point of transition between the old and the new.

Stuart Kauffman and a Home in the Universe

As we've seen, the new finally broke through with the hole poked into the pseudoDarwinian paradigm by chaos and complexity theory—to which I'll return in Book III.

Of the breakthrough books, publication in 1995 of biophysicist Stuart Kauffman's *At Home in the Universe*[22] was particularly attuned to the call for change.[23] With what by now was morphing from chaos into complexity theory, out of Kauffman's book burst a veritable rhapsody of discoveries, terms, and claims for the very old becoming vividly new.

Kauffman's book brought new life to what was uncovered and named by Ilya Prigogine, Francisco Varela, and Vilmos Csanyi, and advanced by Ralph Abraham, Karl Pribram, and Ervin Laszlo among those I knew and worked with in our General Evolution Research Group.[24] Among systems scientists the new thrust had been ecstatically described in 1980 by Eric Jantsch in *The Self-Organizing Universe*[25], then in 1988 touched on by Fritjof Capra in *Uncommon Wisdom*[26] and in 1997 expanded in *The Web of Life*.[27]

More than merely another idea or theory, it was the universally observable fact of the capacity within every organism, including ourselves, to have a voice in the shaping of our future—this, rather than being solely the puppet of forces larger than ourselves.

It was the variation part of Natural Selection and Variation come to life in a powerful new burst of self-recognition. It was the so-called principle of *self-organizing* at work at every level where creativity was in motion throughout the world of humanity and the universe.

To the task of going up against the paradigm to get this fact across Kauffman brought a rare grounding in physics, biology, and the healing orientation of an M.D. in medicine. Through both mathematics and intensive experiments, he demonstrated how in evolution, at a certain level of repetition for a seemingly random process, an inbuilt order within the system takes over and shapes the future into a reasonably intelligible path.

Kauffman further focused on the critical factor even then still missing from almost all this hopeful new work: the prime ingredient for the lost Darwin of moral evolution.

"Then can a heightened consciousness bring about a global ethic?" he asked. "I believe so. I believe, I hope correctly, that what I have sketched ... is true, points to a new vision of our co-creating reality, that it invites precisely an enhancement of our sense of spirituality, reverence, wonder, and responsibility, and can form the basis of a trans-national mythic structure for an emerging global civilization.

"To ever succeed, this new view needs to be soft spoken. You see, we can say, here is reality, is it not worthy of stunned wonder? What more could we want of a God? Yes, we give up a God who intervenes on our behalf. We give up heaven and hell. But we gain ourselves, responsibility, and maturity of spirit. I know that saying that ethics derives from evolution undercuts the authority of God as its source. But do we need such a God now? I think not. Nor do we need the spiritual wasteland that post-modernism has brought us...

"Life is valuable on its own, a wonder of emergence, evolution and creativity. Reality is truly stunning. So if you find this useful, let us go forth, as was said long ago, and invite consideration by others of this new vision of reality. With it, let us recreate spiritual community and membership. Let us go forth. Civilization needs to be changed."[28]

THIRTEEN
THE TWILIGHT OF THE SUPER NEOS

Day gives way to late afternoon. Bird cheeps fade. Here and there one can just barely make out a star. If we may pull back and see the battlefield as if from a helicopter, it becomes evident the whole thing was like an argument under a street lamp that went on for 100 years.

Punctuated by fist fights, it was the thrilling activity at the same street corner, under the same old lamp, to which on hot nights the fans swarmed eagerly to watch. Now up to the plate, or into the ring, dancing, waving hands in fat gloves above his head, dancing while rotating to take in the crowd, steps a fresh new macho champion for the Super Neos.

The crowd cheers. The fight begins. Meanwhile, silent in the shadows stand Jesus, Gautama—or the lost Darwin—unable to be heard. Or elsewhere, simply busy doing what needed doing to *evolve* rather than go backward, were people like Maria Montessori and Martin Luther King.

Meanwhile, what was happening out in the mind space for America and the world at large?

Century after century, journalism, history, and street talk have primarily keyed to whoever were the chief figures empowered by prevailing D-systems. Hence, for the stretch from 1980 into the early 21st century, for the world at large the most impactful story of whether we were evolving or going backward could be collapsed into how the American president of the late 20th century Greed Era set the stage for the American president of the early 21st century ecological, economic, and moral debacle.

Year after year what emerged in policy and action became the great mystery of the time to millions both in America and worldwide. They found it impossible to believe this was actually happening in America. Could this really be the agenda for the land of the free and the home of the brave, of George Washington, Benjamin Franklin, and Abraham Lincoln?

Had the nation been a patient in a psychiatrist's office, a case of trance induced by hypnosis or a brain tumor might have been the diagnosis. But trance on such a wide scale? How was it possible?

But is this rightful to consider in political terms? Isn't politics really out of place? For the bulk of what we were told of, and sold on, evolution theory by the theorists of the 20th century tells us politics is one thing, evolution another.

But is this true? Or is the message that politics is the canary in the mine, the possible clue to the murder mystery, the nightmare that might turn out to be real?

Isn't it evident that everything the traditional theorists of evolution thought of as "just politics," thereby remote from their professional concern, is in reality the central stuff of evolution?

How it is that over and over again the pattern for our history is that of the trance out of which, seemingly all of a sudden, political, economic, or environmental devastation, which either checks or drives us backward, erupts? The Watergate scandal of the Nixon presidency, the Bay of Pigs disaster of the Kennedy presidency, the Iraq war lunacy and economic debacle of the G.W.Bush presidency—the rise and fall of the largest banks and brokerage houses in America, the multi-billion dollar Ponzi schemes, the corruption of mushrooming newcomers like Enron and long time respected brand names for every industry, et cetera, et cetera, et cetera. From child and wife abuse to the big stories of species abuse that rock nations, century after century the trance seems to be the cobra spell that accompanies forays of the devastation of the D-attractor and D-attractor

110

human engineering at work.

And in the case of late 20th century America, what was the purpose of this trance?

Its systems-function seems to have been to mask the construction of a new social, political, and economic edifice dedicated to the 21st century persistence of counter-revolutionary human engineering. But to what end?

If we set aside all the weighty books that have been written to flesh out an answer; if we simply look at what happened with the wisdom of the street, the criteria for the human to be constructed seemed to be something like this.

What was needed was a human who knew how to make money and contribute the huge amounts that were needed for political campaigns. Who knew all the right people and knew how to keep the wrong people down. Who knew how to say all the right things, yet do what was needed and get away with it. Who would obey without question all commands from on high. Who could be counted on for loyalty to the death. But, above all, this—as it was here the corporate sponsors were most concerned. One must construct someone who would automatically buy and buy and buy everything the sponsors built or shipped in from overseas and advertised.[1]

In 1932 Julian Huxley's brother, the novelist Aldous Huxley, in his brilliant dystopia *Brave New World*, first warned of where the D-attractor was driving us. In 1935, in *It Can't Happen Here*, Sinclair Lewis gave this insight a prophetic homespun American setting with the election of Berzelius "Buzz" Windrip, a smiling good-old-boy rightwing jokester for president. In 1949 George Orwell pointed to the ultimate end place for the trajectory of the D-attractor in his book *1984*. Then in 1980, four years after the publication of *The Selfish Gene* and five years after publication of *Sociobiology*, in the year of the election of Ronald Reagan to the first of two terms, a book appeared that got little press, for which sales were minimal, but which seemed to capture an immensity of meaning in just two words.

The book was *Friendly Fascism* by political scientist Bertram Gross. *Friendly Fascism: The New Face of Power in America.*

And what was the relation of the Neos and the Super Neos to all of this? In social scientific fact it cannot be charged they were the cause of our joy ride to a bitter end for the 201th century. Among them were some of the better minds of their time. Among them were men and women of high ideals and unquestionably prosocial intentions. It could be said that what happened to them was a demonstration of the powerful hold upon them and most of the rest of us of the over-riding power of the D-attractor and the Domination system mindset, ideology, and paradigm.

It could be said that this was and is a case of the isolation of the mind space for science within the mind space of America and the world as a whole. The Controverial Connection in the Reflections section ending this book probes this relationship in terms of the super sensitive eugenics issue. As carefully diagramed through the action of attractors within fields in chapters seven, eight, nine, and ten, within the mind space of science the discourse was and is, by necessity, conducted in a host of private languages. But out in the surrounding mind space for everybody else most of us are still so far behind the science of our time that it is as if Freuds and Einsteins have emerged in the Middle Ages.

It is as if the Neos and the Super Neos and their critics argued within a private retreat deep in the woods, or within a sound proof glass room where they could be seen but not heard.

By and large their discourse was discrete and constrained by professional standards. But as it was transmitted by the guesswork of those outside the glass trying to lip read what was being said inside, increasingly it was caught within the backwash of the political, economic, and religious D-attractors for the time.

All this said, the blame for and at times the shame of science cannot be evaded. As, discarding humor, Stephen Jay Gould ventured in the most passionate of his essays,[2] the arrogance of isolation and blind disregard of consequences underlines the crucial responsibility of the scientist and all other privileged scholars for what happens out in the wider world.

Back in Darwin's time the proud catch phrase for Marx was that he "stood Hegel on his head." That is, where the philosopher Hegel maintained that evolution is driven by a dialectic of revolution versus a counter-revolution of ideas, Marx maintained that ideas at the high end were driven by economics and politics at the low end. In this sense of the topsy turvy, The Neos and Super Neos stood Darwin on his head.

Here was Darwin, attuned to thousands of years of progressive theology and hundreds of years of progressive philosophy, right side up in maintaining that as we developed the moral sense became paramount in driving human evolution. And here were the Neos and the Super Neos, attuned to an update for the doctrine of Original Sin as prime driver. So they flipped him upside down, and out in the wide world when the question came up of "What is evolution," for lack of anything better, up came the handy answer provided by the mind-mangling magic of the D-attractor and domination system.

If you had graduated from high school, the answer was "survival of the fittest." If you had finished college and perhaps gone on to graduate study, one might add it was something to do with "selfish genes." And thereafter, across all the levels for human activities for our species, this mindset of "survival of the fittest" and the "selfish gene" significantly shaped the action for both the people and nations of the 20th century.[3]

As is said of those who with the best of intentions are the enablers of everything from alcoholism to drug addiction, the Neos and Super Neos became enablers of the greater sickness of the over-riding paradigm.

113

Though cloaked in the robes of academia and wearing the most gloriously tasseled mortarboards on great occasions, they became the scientific Typhoid Marys of their time.

But what of the fact that most of the Neos and Super Neos were self-avowed liberals, even progressives? Why would they blindly go on year after year ignoring how their work was being used to pave the way to the D-attractor future?

Under increasingly biting criticism some later shifted to the mind and language of reform,[4] but why did so many still persist and prosper, as if none of this had ever happened?

This question I must leave to psychologists far more gifted than I am. I can only provide food for thought in Book II.

The Super Neo Bible

Of the many books produced to advance the Super Neo case for sociobiology's offspring, evolutionary psychology, early on one in particular became known as their bible.

Edited by anthropologists Jerome Barkow and John Tooby and cognitive psychologist Leda Cosmides, published in 1992, in *The Adapted Mind: Evolutionary Psychology and the Generation of Culture* twenty-five bona fide social scientists set out to prove theory with active intervention in a world of real immediate needs—this, rather than only speculatively on paper, as could be said of Stuart Kauffman and other theorists for the opposition.

In surprising ways, *The Adapted Mind* was the statement of an agenda for proving theory through action research that called to mind the pioneering work of Kurt Lewin and his students that transformed psychology a half century earlier.[5] Out from the sociobiological base they pushed into such badly needed areas for an evolution-oriented activist psychology as how can we shift from gut-cutting competition to cooperation, how best improve mating and sex, parental care and children, how do perception, language, and intrapsychic processes fit into the picture for action, even the intriguing, all-too-seldom-come-by idea of how do we improve environmental aesthetics.

In many ways a 136 page opening chapter "The Psychological Foundations of Culture" by John Tooby and Leda Cosmides was an impressive intellectual achievement. It was also an important historical document for what it recorded of the legitimate strengths and prospects for this fresh new contender.

On the positive side, Tooby and Cosmides made the same critique of social science that many others have tried to advance: of the weakness of a social science divorced from natural science. How chopped up into all the little divisions that ignore one another, social science is like a throwback to the age of feudalism, with each little barony guarding its territory against the others, ungrounded in either the perspective of or loyalty to evolution or evolution theory.

Their critique of a traditional social scientific bias against the idea of anything being innate or "biological" in human behavior was also well founded. Their stress on empirical rigor and a careful attempt to link the findings of chemistry, biology, information science, etc., with cognitive psychology was similarly well-grounded.

Barring the problems we'll come to, their critique of social science was, in short, the kind of historically needful corrective—or kick in the pants—that social science needed.

But then came the slide down the slippery slope of what became the Super Neo mentality into the old blood sport for the D-attractor's captive cohort. Setting up the great sociologist Emile Durkheim and, more recently, the revered anthropologist Clifford Geertz, as straw men for the villains of

115

their piece, they painted a woefully cock-eyed picture of 20th century social science as almost wholly based on the idea—which in fact went out of science centuries ago—of the mind as a "tabula rasa" or "blank slate."

This was the phrase to which the fresh new champion for the Super Neos, psychologist Stephen Pinker, returned in his best seller *The Blank Slate*. Social science, Tooby and Cosmides claimed, looks at the organism —that is, us along with all other creatures—as an "empty vessel" to be filled by the astonishment they fixed on for the central villain.

For the scoundrel of their manifesto for human salvation was none other than the process of *learning*! The monster was this process upon which our species has presumably been foolishly expending time, energy, and an incalculably huge investment in schools for at least 5,000 years.

This is such an astounding claim we must take a close look at how they arrived at it.

Social science, they tell us, has been overwhelmingly governed by the idea of "nurture" to the exclusion of "nature." In other words, a large part of our problem today is that we have presumably become the unthinking captives of the idea that everything we are is governed by what happens to us as we go through life rather than by what we are born with through the transmission of genes and the dictates of biology.

The simple historical and unquestionable fact is that the hoary "nature versus nurture debate," which bogged down psychology in the early part of the 20th century, was long ago put way behind us.

To put this within the picture for a new language for evolution, so prevalent across all fields involved was the knowledge that we are composed of nature *plus* nurture that by the late 1940s Kurt Lewin could put it in the form of a simple formula $B = f(P, E)$. Or **B**ehavior is a function of **P**erson and **E**nvironment.

The perception of what is supposedly to fully function as a human being can be glimpsed through passages such as this.

"Just as the fields of electrical and mechanical engineering summarize our knowledge of principles that govern the design of human-built machines, the field of evolutionary biology summarizes our knowledge of the engineering principles that govern the design of organisms, which can be thought of as machines built by the evolutionary process."[6]

One could pass this by as only a metaphor, except for two facts. One is that the suggestion in context is that we are little more than machines built by the evolutionary machinations of some Great Robot in the Sky. The other is that this rather horribly smacks of D-attractor construction of a handbook for late 20th century human engineering.

Worth carving in stone over the portal of every school of the 21st century are the names of the psychologists and neurologists involved in the study of perception, motivation, memory, and language, whose work provides the foundation for the field of education and at least 300 years of gaining a reasonably good understanding of the process of learning.

For example: William James, Ebbinghaus, Dewey, Montessori, Piaget, Thorndike, Terman, Pavlov, Bechterev, Watson, Skinner, Koehler, Wertheimer, Guthrie, Hull, Lashley, Hilgard, Dollard, Underwood, Milner, Mednick, Harlow, Hebb, Lashley, Bartlett, Guilford, and Pribram and MacLean in brain research.[7]

Yet of all this effort Tooby and Cosmides in effect tell us that well-meaning souls such as these merely "reified this unknown functionality, imagining it to be a unitary process and called it 'learning.'"[8] This concept, they tell us, not only remains "in genuine need of explanation," but "will eventually disappear as cognitive psychologists and other researchers make progress in determining" what is really going on here.[9]

They tell us that the same progressive oblivion lies ahead for such

117

concepts as "culture," "intelligence," and "rationality."

Moreover, within the 666 page mass of *The Adapted Mind*, which purports to map the future for us, among hundreds of references and thousands of index entries, only a single item refers to a single page having to do with anything about moral sensitivity, or moral development, or morality.

And when one goes to this single item what does one find? Typically some post-modernist academic mush under the revealing subhead "Conscience, Guilt, and Neurosis."

Under the bold banner of evolutionary psychology is this to be the brave new world they herald?

Is this supposedly what's to come when the delusions of the thousands of social scientists and educators who came before this spunky band of scientific salvationists are, as they trumpeted, "replaced with knowledge"?[10]

Aren't we here looking at the question of whose science, whose politics, whose economics, whose education, whose technology, and whose religion is to win the battle for 21st century mind and human survival?

EPILOGUE
YESTERDAY, TODAY, TOMORROW

I'm told I've been unfair and needlessly divisive in my critique of the Super Neos. I'm told the hard edges are being softened and the field is maturing.

Let us hope history confirms this. The fact remains that in this book I document the hard evidence of books published within and their impacts on the 20th century. However the Super Neos may work to bury or otherwise gloss over what they originally wrote, it must serve as a grim warning to the 21st century. For the gyrations of evolution show us that as inevitably as night follows day, after a lull their heirs will return.

I'm also asked why I haven't added more new works being heralded since the turn of the century to advance the dialogue on the theory and story of evolution.

Surely to recap the theory and tell the story of the aspiration and the damage of a single century is enough, hence this book is almost wholly confined to books published within the 20th century—albeit with impact for some, and let us hope for the best rather than the worst, for decades yet to come. For as with the virus that hangs on within us by changing its shape but not its end, the challenge I address in this book will not go away for a very long time.

In a way what happened in the 20th century goes back to an observation by Darwin in *Origin of Species*, probably skipped past for well over 100 years as being too simplistic. After all, it's the complexities, which one can battle over for decades before reaching any consensus, that generate the debates, books, and everything else that advance a career in politics, religion, publishing, and science.

"The most important of all causes of organic change," Darwin tells us in *Origin of the Species*, "is one which is almost independent of altered and perhaps suddenly altered physical conditions, namely the mutual relation of organism to organism,—the improvement of one organism entailing the improvement or extermination of others."[1]

Improvement through *improvement* rather than extermination of others?

This seems too quiet, too simple to make much of. Yet from the point in time at which Darwin wrote this we can see yesterday, today, and tomorrow take shape far into the future.

Darwin focused on improvement through the extermination of others in *Origin* and then on improvement through the improvement of others in *Descent*.

Like the flow of two rivers from springs nearby one another in the deep past, the history and the next stage for the evolution of our species unfolds. One stream, of necessity polluted at the outset but encouraged in every way by the Powers That Be, accelerates evolution through the elimination of others.

The other stream, of the Powers that *Should* Be, although continually blocked and diverted by the Powers That Be, still clear and potable, pushes on to advance evolution through the improvement of others.

Within the fields of science and religion and the realities of politics and economics the difference between the two streams sharpens into warring baronies along the polluted stream and peaceful villages along the clean stream. Over centuries this difference has hardened into the battle between a higher order and a lower order alignment for control of 20th and 21st century mind—and thereby human survival.[2]

At issue in the battle is whether we are to be guided:

• *socially*, by a theory and story of subhuman, inhuman, and inhumane evolution, or a theory and story of a *fully human* evolution.

• *politically*, by a theory and story of adaptation and accommodation, or a theory and story of assertion and aspiration.

• *economically*, by a theory and story of selfishness and greed as the key to wealth and power for the privileged—with, of course, a noble trickle down to the many. Or a theory and story of caring and sharing to spread the wealth of Earth to the many.

• *scientifically*, by a reductionist theory and story oriented to the deep past, applied spuriously to the present, or by an expanding theory and story oriented to today and the future.

• *morally*, by a theory and story of survival of the fittest and selfishness *uber alles* as the driver for human evolution, or a theory and story of love and moral sensitivity as the driver of human evolution.

• *spiritually*, by a theory and story of original sin and a ferocious God of my people versus your people, a God that must without question be obeyed. Or a theory and story of original blessing and a merciful God of all people. Or no God at all, good will from on high and below, just favoring an abundant life for all.

The good news is that through the escalation of disaster driven by the wrong choice we're being forced to grow up and move out of the paranoid mind space that requires a battle over every step of the way.

Gradually, although by no means fast enough, the science of the old choice and the wrong choice is being modified or replaced by the evolutionary systems science growing out of humanistic social science and an enlightened new biology shaped by cross-disciplinary cybernetic, chaos, complexity, and self-organizing theory. Even so, within religion, politics, and economics the resistance to sanity is still obdurate, all three like a spoiled child resisting every attempt by progressive religion, progressive politics, progressive economics, and just plain sanity to advance our species.

Above all, within all fields there is movement toward recognition of evolution itself as the point of both origin and destination.

What happened will happen again until the lopsided theory and story that drove us to the edge of a cliff in evolution has been replaced with the grown-up theory and story that was Darwin's and now is his progressive successors' vision.

As Darwin wrote, it is the "*relationship* of organism to organism" that matters—that is, as all the great spiritual visionaries have said, it is what we do or do not do unto one another.

For thousands of years it has been so simple in the end.

REFLECTIONS AND AN INVITATION

ROMANES ON DARWIN'S LEGACY

Beyond the brief sketch of the agony of Romanes in chapter one lies much more to the haunting prophecy of what lay ahead for the course of what Romanes was first to call NeoDarwinism.

As he lay there dying of a brain tumor, and the doctors came and went and everyone tried to look on the bright side, in the manuscript for what became his "lost" book, *Darwin and After Darwin*, Romanes hammered away at trying to get across what was at stake. The writing of "his" Darwin out of mind by misguided successors must not endure.

"The question is whether natural selection has been the sole, or but the main, cause of organic evolution," he wrote to establish the grounding point for his case for Darwin's real legacy, versus what was already being degraded into the Social Darwinism of "survival of the fittest" by the Robber Barons of that time.[1]

Because of the difficulty most of us face in grasping the differences and consequences posed by what often seem awfully picky points for the evolutionists, he labored at length to get his message across.

"Of all the errors connected with the theory of natural selection, perhaps the one most frequently met with—especially among supporters of the theory—is that of employing the theory to explain all cases ... The term 'natural selection' thus becomes a magic word, or Sesame, at the utterance of which every closed door is supposed to be immediately opened."[2]

"I am not here alluding to that merely blind faith in natural selection, which of late years has begun dogmatically to force this principle as the sole cause of organic evolution in every case ... Such a blind faith, indeed, I hold to be highly inimical, not only to the progress of biological science, but even to the true interests of the natural selection theory itself ... The point is, that the theory in question is often invoked in cases where it is not even logically possible that it can apply, and therefore in cases where its application betokens, not merely an error of judgment or extravagance of dogmatism, but a fallacy of reasoning."

This shows, he notes, "how easy it is to be cheated by this mere juggling with a phrase."[3]

Thereafter, Romanes focused on areas of concern that all too often 20th century science and education was either to fumble, miserably mangle, evade, or as if it was something wholly new born from the forehead of a new Jove, to rediscover and hail late in the century.

The effect of education

To pound across this soon-to-be ignored factor for Darwin's expanding perspective on evolution, Romanes notes "the unquestionable influences of individual education."[4]

This seems obvious to us, but ahead, for a whole century, lay the battle between those holding this belief and those who were to claim that education had little effect, that only the quality of genes mattered, that it was mostly nature with minimal nurture involved.

Does this matter today?

Crossing over from scientific mind to political and economic mind, by the late 20th century this became the battle for vouchers of public money to support private schools. Underneath more acceptable rationales was the simple subliminal message. Public money to private schools would guarantee transmission of the gene pool of good money makers and good

Christians to counter—God forbid!—the threat to human evolution of public school transmission of the gene pool of bad races and classes.

Surely, Romanes wrote, it was "sufficiently well known that individual education—or special training, whether of mind or body—usually raises congenital powers of any kind to a more or less considerable level above those of the normal type. In other words, whatever doubt there may be touching the *inherited* effects of use, there can be no question touching the immense *developmental* effects thereof in the individual life-time."[5]

Thereafter, Romanes tackled the issues in which, though dying, he enjoyed an immense advantage over most of his contemporaries and peers in science.

Almost all of them were beholden to sources of established wealth for employment—prestigious colleges and universities, foundations, or other organizations, all of which looked askance on those who were in any way labeled controversial. They were further dependent on the acceptance rather than the blackballing of professional associations. Beyond all this their time for independent thought and work was radically diminished by the long hours one must spend in teaching and in the writing of books and papers that might safely boost rather than risk their advancement.

For Romanes, however, this was no problem. He had Darwin's great advantage. Having inherited a fortune, he didn't need to teach or do anything else but work for himself. He could set his own agenda, finance his own research, invest with gifts in the work of others whose work he admired and believed in.

Like Darwin before him, here again in his disciple emerged the rare independence that made it possible to go up against the paradigm. Consequently, one after another, he was able to tackle issues over which the battle still rages on into the 21st century.

David Loye

The idea that selfishness and survival of the fittest are the prime drivers of evolution

"It constitutes no part of the theory of natural selection to suppose that survival of the fittest must invariably lead to *improvement* of type ... Parasites, for example, present the phenonmena of what is called *degeneration*, i.e., showing by their whole structure that they have descended from a possibly very much higher type of organization than that which they now exhibit."[6]

Cooperation versus every man for himself

Even "members of different species mutually assist one another" in "what is called symbiosis," Romanes wrote.[7]

Referring to the early appearance of the articles that a decade later Prince Peter Kropotkin brought together in his book *Mutual Aid*, Romanes tells us that "all cases of mutual aid or co-operation, whether within or beyond the limits of species ... fall under the explanatory sweep of the Darwinian theory."[8]

The tunnel vision of biology

Another observation by Romanes gains in significance when we consider the 20th century consequences not only of ignoring Darwin's completion of theory, but also of the virtual exclusion of the social sciences from the formal body of evolution theory by a monopoly of biology.

"As there is no necessary relation between erudition in one department of science and soundness of judgement in another, the mere fact that a man has distinguished himself as a botanist or zoologist does not in itself qualify him as a critic where specially Darwinian questions are concerned," he wrote.

"It was Darwin's complaint that for many years nearly all his scientific critics either could not, or would not, understand what he had written—and this even as regarded the fundamental principles of his theory, which with the utmost clearness he had over and over again repeated."[9]

Sexual selection

A glaring instance of the exclusion of whatever didn't fit the early agenda for the Neo-Darwinians was Darwin's pet theory of the impact of sexual selection as well as natural selection in shaping all life on this planet.

There "can be no question that the courtship of birds is a highly elaborate business," Romanes wrote, "the inference is that the males do not take all this trouble for nothing; but that the females give their consent to pair with the males whose personal appearance, or whose voice, proves to be most attractive."[10]

The "evidence in favor" of the operation of sexual selection, Romanes tells us, "is both large in amount and massive in weight." Yet not until relatively late in the 20th century—with the surge of power for the women's movement, and with an eye to its obvious book marketing appeal for males as well as females—did the Super Neos morphed from the Neo-Darwinians turn to the topic as if it were via them newborn.

Creation by design and religion

Here Romanes stated what among the evolutionists all sides to the battle—as well as an overwhelming majority of people of any degree of liberated mind—were to wearily unite on for well over a century.

If all the creatures on this earth including ourselves "came into existence suddenly, the fact would constitute well-nigh positive proof in favour of ... creation by design; whereas, if they all came into existence gradually, this would in itself constitute presumptive evidence in favour of naturalism, or of development by natural causes."[11]

So does this mean the Creationists and Evolutionists must go on until the earth burns to a crisp wasting our precious time in senseless argument?

Out of the feeling that was clearly also Darwin's and his own increasing conviction of no real barrier between progressive science and progressive religion, Romanes ends the first volume of *Darwin and After Darwin* by noting that "the religious thought of our generation has been more than ever staggered by the question—Where is now thy God?"[12]

A fascinating and promising phenomenon for the early years of the 21st century is the emergence of a new meeting ground for progressive science and progressive spirituality in a mutually reinforcing alignment to the cause of evolution.[13]

In keeping with the *Zeitgeist* of an interface of toleration for belief in God or no God is Romanes' discrete response.

"The logical standing of the case has not been materially changed." When the "cry of Reason pierces the heart of Faith, it remains for Faith to answer now, as she has always answered before—and answered with that trust which is at once her beauty and her life—Verily thou art a God that hidest thyself."[14]

An embryo for chaos theory?

Further underscoring Darwin's early perception of the so-called butterfly effect for chaos theory in what he called *correlated variation* also engaged Romanes.

"Mr. Darwin, who has paid more attention to this matter than any other writer," Romanes affirms, "has shown, in considerable detail, that all the parts of any given organism are so intimately bound together, or so mutually dependent upon each other, that when one part is caused to change by means of natural selection, some other parts are very likely to undergo modification as a consequence ... It is evident that in this principle we find a conceivable explanation of the origin of such adaptive structures as could not have been originated by natural selection..."[15]

THE BATTLE FOR HEART AND
MUCH LARGER MIND

In this first book for our trilogy we've looked at the struggle that many would agree was the most productive, dramatic, and important of battles for 20th century science. From the perspective of the 21st century, however, another stream of thought is vital. Long dominant in Eastern religion, it was foreshadowed in Western science at least as far back as 1930 by the remark of astronomer Sir James Jeans that "The universe begins to look more like a great thought than a great machine."[1]

In other words, over the centuries both in Western science and everyday perspective, the focus has been on the rise of mind out of matter. But the discourse is shifting back to where it used to be in philosophy, to the rise of matter out of mind. Emerging out of an expanding interface between science and spirituality is the popular spread of "New Age" works by Ken Wilber,[2] Ervin Laszlo,[3] Christian de Quincey,[4] Michael Lerner,[5] and many others to which I will turn in Book III: *Up Against the Paradigm*.

Here looms still another dimension to the battle for human survival. For here again can be seen the clash of revolutionaries pushing for progressive freedom, equality, and *expansion* of mind and paradigm, and counter-revolutionaries pushing for regressive control, inequality, and *diminishing* of mind and paradigm.

In *The Evolutionary Outrider: The Impact of the Human Agent on Evolution*, in 1998 I introduced the concept of the evolutionary outrider.[6] The term was meant to capture the nature and situation of those who boldly scout on ahead of us to explore potentially revolutionary as well as evolutionary prospects for a better future.

These evolutionary outriders, I see now, are of two kinds. One set is of the 20th century theorists, whose story we've tracked, who seek to better understand, and teach, and thereby help guide evolution scientifically or spiritually.

The other group, however—orienting to the lost Darwinian half emphasizing love and moral sensitivity—simply set aside the often alien tangle of theory to listen to their hearts, then go ahead and work to advance evolution along paths that seem obvious to them into the future.   I t might seem that from this relationship one might ask the question often raised by those making the case for action versus theory. Why bother with the fusty delay of theory? Why not just go ahead and get done what needs to be done?

This takes us to the open pit of the great hole in modern mind: how for lack of a sufficiently updated, expanded, and unquestionably *progressive* theory of evolution both kinds of evolutionary outriders were checked in place, sidelined, or driven backward by D-systems fixation on the first half degradation of "survival of the fittest" and "selfish genes" for Darwin's purported theory.

In the dynamics of the D-attractor and D-system can be seen the power of the over-riding, transnational, political, economic, and religious mindset functioning to either hold us in place or drive us backward in evolution. In terms of another language for evolution can be seen the metaphorical manipulation of the invisible puppet master, whereby thousands of scientists were and are, like children, penned in to play in a giant sandbox while the "grownups" go off to rule the world.

This is vital to keep in mind as we turn to a brief look at the potential for both theory and action if they can move ahead together.[7]

What's possible was foreshadowed in 1978. In that year, in an old house in Princeton all covered with vines, three evolutionary outriders gathered for a week-long brainstorming house party.

There was no great fanfare or audience, just the three of them, three women who loved and admired each other, with big plans for the future.

The three were Hazel Henderson, with significant impact lying ahead on the fields of futures studies, a major critique of economics, and a global television program advancing the cause of an ethical marketplace; Jean Houston, who became a charismatic founder of the human potentials movement; and Barbara Marx Hubbard, who in evolving herself became a visionary driver of the new field of conscious evolution.

Their rollicking talks were tape recorded and a transcript made but never published. As so often happens in these stories, however, thirty years later the transcript was dug out of the proverbial attic and because of their prophetic exploration of issues that have come to matter in our time was published as *The Power of Yin*.[8]

Here's a sense of how these evolutionary outriders explored their sense in common of a mission and paths opening into the better future.

Hazel "We are, all three of us, trying to play midwife to evolutionary growth processes, and in so doing we subject ourselves to almost intolerable levels of stress ... I've accepted the fact that the stress is going to be there until I die because I'm doing what I ought to be doing. I'm throwing myself into the stream of the process—and it's okay."[9]

Jean "What I learned from our experience was the power of creative loving and of the mutual eliciting that takes place at that level of communality ... And this is something that—at least at this point in time—seems to be much more possible among women ... I am a member of many top-heavy, hierarchical organizations —many of which I sit on top of! To come into a situation in which there is so much love, so much sharing, listening and mutual eliciting—and laughter!—was probably the most democratizing experience of my life!"[10]

Barbara: "We have reached 'critical mess.' A perfect crisis!... Evolutionary women are on the rise. And we seem to create a new context in which evolutionary men are freer to release themselves from the bonds of patriarchy and join together in new forms of real partnership. We are giving birth to a new human and a new humanity within ourselves."[11]

Same year, 3,000 miles away on the West coast in Los Angeles, a deep-thinking activist rising out of the women's movement had just met a man to "release from the bonds of patriarchy" so together they might explore "new forms of real partnership." The man was myself. Ahead for Riane Eisler were major books on cultural evolution, sex, politics, economics, education, and spirituality, published in 24 languages, on the global impact of partnership systems, or gylany, versus domination systems, or androcracy.[12]

"For above all, this gylanic world will be a world where the minds of children—both girls and boys—will no longer be fettered," she wrote of her vision of the better world in *The Chalice and the Blade*. "It will be a world where limitation and fear will no longer be systematically taught us through myths about how inevitably evil and perverse we humans are. In this world, children will not be taught epics about men who are honored for being violent or fairy tales about children who are lost in frightful woods where women are malevolent witches. They will be taught new myths, epics, and stories in which human beings are good; men are peaceful; and the power of creativity and love—symbolized by the sacred Chalice, the holy vessel of life—is the governing principle. For in this gylanic world, our drive for justice, equality, and freedom, our thirst for knowledge and spiritual illumination, and our yearning for love and beauty will at last be freed. And after the bloody detour of androcratic history, both women and men will at last find out what being human can mean."[13]

The year was 1999, and out of the entertainment world stepped a uniquely placed voice to further the case for children. Seldom formally noted has been the impact of music, and more widely entertainment, on the course of history that shapes the course of evolution. Beethoven and Mozart, Black spirituals, the Beatles, the thousands of composers and performers of songs of protest, love, and the joy of living have had an inestimable power in undermining the wrong Powers That Be and the strident voices of hate.

Child himself of survivors of the Turkish massacre of Armenians, over decades Raffi had become chiefly known worldwide as the beloved composer, performer, and singer of songs for children. But the other side to Raffi was an increasingly fervent involvement in environmental and social action.

"We find these joys to be self-evident," he wrote in a Covenant for Honouring Children backed with the push of his later formation of the Canadian Centre for Child Honouring. "That all children are created whole, endowed with innate intelligence, with dignity and wonder, worthy of respect. The embodiment of life, liberty, and happiness, children are original blessings, here to learn their own song. Every girl and boy is entitled to love, to dream, and to belong to a loving 'village.' And to pursue a life of purpose," he wrote in a paraphrase of Jefferson's Declaration of Independence.

The year was 1993. The Vatican had forbidden the teaching of progressive theology by one of the greatest living theologians. Now the German heretic, Hans Kung, waited to see what was going to happen to his draft for a Global Ethic during the once-every-decade historic meeting in

Chicago of the Parliament of the World's Religions.

Reception for his proposal for a moral code in common worldwide was a big question mark, for it was a daring statement of P-attractor principle sure to run smack up against the global entrenchment of the D-attractor.

"We are women and men who have embraced the precepts and practices of the world's religions," emerged out of the pages scanned by the throng—no doubt provoking shock, even possibly outrage, within many Imams and other defenders of the patriarchal paradigm. Women and men? Just a minute! Really now!

"We affirm that there is an irrevocable, unconditional norm for all areas of life, for families and communities, for races, nations, and religions...We must treat others as we wish others to treat us ...We consider humankind a family ... There should be equal partnership between men and women." (There, God help us, it is again!). "We commit ourselves to a culture of non-violence, respect, justice, and peace ...We must move beyond the dominance of greed for power, prestige, money, and consumption to make a just and peaceful world."[14]

Kung's new Global Ethic was of course longer than this sample. Back and forth for months, he'd wrestled with 200 other religious scholars on the draft. Could it possibly pass? It did pass!

And signed not only by a grudging handful. 143 respected leaders from all the world's major faiths had signed to put their stamp on an historic move ahead in evolution—Baha'i, Brahmanism, Brahma Kumaris, Buddhism, Christianity, Hinduism, Indigenous, Interfaith, Islam, Jainism, Judaism, Native American, Neo-Pagan, Sikhism, Taoism, Theosophist, Unitarian Universalist and Zoroastrian had signed!

The year was 1984. While somehow also managing the miracle of raising ten children, Stanford professor Nel Noddings had just published *Caring: A Feminine Approach to Ethics and Moral Education*.[15] Behind her was the tradition of a morally-oriented progressive education going back to Pestalozzi in 1801. Maria Montessori and John Dewey were among her treasured predecessors. Currently, Ron Miller, Tim Seldin, and others were probing the prospects for evolutionary advance in this direction.

Particularly in keeping with Darwin's and later Piaget's emphasis, Noddings had forged ahead to found a nationally influential perspective on the intimacy of family life and parent-child relations in grounding the impact of education on evolution.

"We should want more from our educational efforts than adequate academic achievement," she urged. "Caring parents and teachers provide the conditions in which it is possible and attractive for children to respond as carers to others ... Children educated in this way gradually build an ethical ideal, a dependable caring self. A society composed of people capable of caring—people who habitually draw on a well-established ideal—will move toward social policies consonant with an ethical care."[16]

The year was 1973. An unknown 24-year-old from Oklahoma City who seemed to others to be into some pretty weird stuff had just completed writing his first book. After rejection by twenty publishers, *The Spectrum of Consciousness*[17] was finally accepted and published in 1977 and Ken Wilber was on his way. Self-taught in Western philosophy, psychology, Buddhism, the Tao te Ching, and the evolutionary thought of Sri Aurobindo, Wilber soon became a chief attractor for the enormously popular global melding of Eastern and Western wisdom that swept out of the 20th into the 21st century.

"Are the mystics and sages insane?" he wrote to pose the conflict over what was evolution and where it was or wasn't taking us. "The story of awakening one morning and discovering you are one with the All, in a timeless and eternal and infinite fashion. Yes, maybe they are crazy, these divine fools ... But then, I wonder. Maybe the evolutionary sequence really is from matter to body to mind to soul to spirit, each transcending and including, each with a greater depth and greater consciousness and wider embrace. And in the highest reaches of evolution, maybe, just maybe, an individual's consciousness does indeed touch infinity—a total embrace of the entire Kosmos—a Kosmic consciousness that is Spirit awakened to its own true nature. It's at least plausible. And tell me: is that story, sung by mystics and sages the world over, any crazier than the scientific materialism story, which is that the entire sequence is a tale told by an idiot, full of sound and fury, signifying absolutely nothing?"[18]

The year was 1984. One night Stanley Krippner had a dream. His friend the native American healer Rolling Thunder suddenly appeared looking sad. There was a coffin beside him. Then Krippner heard the voice of Rolling Thunder's wife, Spotted Fawn. "You know, I won't be seeing you any more."[19] On returning home he found that his beloved friend Spotted Fawn had died that very night.

Picking up from where we've seen Alfred Wallace and William James were early in the century, this was typical for Krippner's other life. While half of him became the highly respected president of the Association for Humanistic Psychology engaged in teaching and the activities of dozens of other formal professional bodies, the other half roamed America, Russia, and South America exploring the world traditionally forbidden to science of the paranormal.

First out of physics, and then more slowly out of psychology, steadily it became apparent that here was a vital frontier for expanding the exploration of evolution. Krippner warned of charlatans and frauds, but year after year stood firm for what he found to be true.

As another explorer of this ancient new frontier, Dean Radin, put it, the implications "are sufficiently remote from engrained ways of thinking that the first reaction to this work will be confidence that it's wrong. The second reaction will be horror that it may be right. The third will be reassurance that it's obvious."[20]

The year was 1969. The 21-year-old heir to the fortune of the founder of the globally expanding Baskin-Robbins Ice Cream company was caught on the horns of an immense dilemma. Should he follow in his father's footsteps with the ice cream business and ultimately inherit the fortune, or follow his heart and lose it all?

Already he had a wife to support. The fortune could guarantee them and eventually their children and grandchildren with security long into the future. But something, he didn't know what it was, only that it was larger, called with a voice that couldn't be denied.

He left the ice cream business to move with his wife Deo to a shack on an island. There they lived a bare subsistence but happy life raising their own food and child Ocean while sorting out what life was all about and what they could best do to help make it better for everybody.

Out of this experience, in book after book—*Diet for a New America, The Food Revolution, Healthy at 100*—came John Robbins' impact with what historians are likely to conclude was the most successful of 20th century revolutions.

Having sorted out what he might do best with his life, John's conclusion was that he left for the island to "pursue the deeper American Dream ... the dream of a society at peace with its conscience because it respects and lives in harmony with all life forms. A dream of a society that is truly healthy, practicing a wise and compassionate stewardship of a balanced ecosystem."[21]

The year was 1980. A doctor on the staff of the Harvard Medical School faced a difficult decision. On one hand she'd come from Australia to bring her expertise in treating children with cystic fibrosis to America. On the other hand, she was convinced the children of the whole world faced the worst scale of disaster imaginable if nothing could be done to stop the insane race between the U.S. and Russia to build and aim at each other more and more thousands of hydrogen bombs on hair trigger alert.

Deciding the situation in the U.S. was hopeless during the astronomical military build-up of the Reagan years, Helen Caldicott gave up her career and returned to Australia. In 1971 she'd led the Australian opposition to French atmospheric nuclear testing in the Pacific. Now she set out to raise hell worldwide.

She co-founded the 23,000 member Physicians for Social Responsibility and the Women's Action for Nuclear Disarmament (WAND) in the U.S. She traveled the world to help start similar medical organizations in many other countries, like the International Physicians for the Prevention of Nuclear War, which won the Nobel Peace Prize in 1985.

She was widely condemned for comparing Mikhail Gorbachev to Jesus Christ in ending the Cold War and for her "distortions" and use of "extreme language"—for example, "As it is, life in America amounts to a corporate dictatorship." Others, however, found hers to be the welcome fire of a heroic voice .

"An aroused citizenry can still move its government to the side of morality and common sense," she wrote in *Nuclear Madness*. "In fact, the momentum for movement in this direction can only originate in the heart and mind of the individual citizen. Moreover, it takes only one person to initiate the process, and that person may be politically naive and inexperienced, just as I was when I first spoke out."[22]

The year was 1987. A tall, slim black man stood in a cemetery between the graves of the father he knew only briefly before being abandoned and the grandfather he never knew. Back sixteen years went his mind to memories of that father; and then, and thereafter, ahead to probe the mystery and the question of how the world evolves from father to son to father again.

The only significant time he'd ever had with his father was a single month in Chicago when he was ten. He'd soared with the embrace of this brilliant handsome man with a deep rumbling voice. They'd danced together at a Dave Brubeck concert. Then swiftly the bubble burst and he withdrew into the protective shell that is the consolation of the abandoned.

Now as he stood there between the two graves in Kenya the memory of the excitement of that visit came back. Then transcending the hollow place within him came a sense of the rare capacity and the power of the ambition within his father to do great things for Kenya. Seemingly come to nothing. Then within the stillness came the sense of the empowerment of the son, himself, to fulfill what had earlier been lost or denied.

"When my tears were finally spent, I felt a calmness wash over me," Barack Obama wrote in *Dreams from My Father*. "I felt the circle finally close. I realized that who I was, what I cared about, was no longer just a matter of intellect or obligation, no longer a construct of words. I saw that my life in America—the black life, the white life, the sense of abandonment I'd felt as a boy, the frustration and hope I'd witnessed in Chicago—all of it was connected with this small plot of earth an ocean away, connected by more than the accident of a name or the color of my skin. The pain I felt was my father's pain."[23]

The year was 1992. Still young, pretty, yet already a long time member of the German Parliament, Petra Kelly was at the height of her global fame as the courageous founder of the new Greens Party fighting for peace and the environment. She was at home in their apartment with her much older, long-time lover.

Beneath her facade of charismatic courage lurked the anxieties of a fearful child. Behind his switch to the Greens Party was the ambivalence of an ex-captive of the D-attractor—World War II fighting for the Nazis, extreme right wing views, an ex-general and now a classic case of the authoritarian personality subjected to what her perceived as domination by a woman. All of a sudden he shot her in the head, then killed himself.

This now we can see was no mystery. Like a majority of the cases in a psychiatrist's office, like a majority of those in prisons worldwide, like a majority of the children starving, the women raped, the civilians and the soldiers slaughtered in all our wars, both Petra Kelly and Gert Bastian were victims of the trampling of the D-system paradigm, mindset, and ideology on the flowering of the health of the P-system, which ever faster than it can be crushed now again rises.

"We, the generation that faces the next century," Petra Kelly wrote before she died, "can add the solemn injunction 'If we don't do the impossible, we shall be faced with the unthinkable.'"[24]

Given the evidence of thousands of evolutionary outriders such as these, how on earth could anyone have ever thought the prime drivers for human evolution are "survival of the fittest" and "selfish genes?"

Consider what could happen if the potential outriders among millions of us were given a theory and story of evolution that felt tailored to them like a good suit of clothes, rather than a bad fit to be avoided, or an embarrassment they were ashamed to wear.

THE CONTROVERSIAL CONNECTION

There were three parties to the eugenics fracas. First to pounce were the creationists.

Seeing in the shock value of Hitler's global devastation a new weapon for their attack on science, the creationists set out to blame Darwin, evolution theory, and thereby science as a whole for Hitler and the gas chamber brand of eugenics.

This, they screamed, was the evil incarnate proof of the collapse of morality under the onslaught of a Godless Darwinism and its bloody theory.[1]

Outraged, seeing in the charge only fresh proof of the colossal ignorance of the creationists, the scientific community closed ranks in common cause with the Darwinians and Darwinian theory. How on earth could anyone in their right mind possibly link the nobility of Darwin to the depravity of Hitler.[2]

Emerging within the scientific community, and more generally the humanities, however, was a minority troubled by the charge. To the glee of the creationists and the dismay of the evolutionists, their concern was openly voiced by the most popular Darwinian of his time, beloved essayist and multiple book author Stephen Jay Gould.

Gould had no use for the creationists, deploring their dismal attempt to drive us backward in evolution. But in a quietly persisting handful of essays and books, he tracked the influence of the survival of the fittest social interpretation of Darwinian theory on public policy—including the jolt of Amoneering of the kind of eugenics that exploded into the ultimate outrage to humanity of Hitler and fascism more generally.[3]

It remains a sore point to this day, but it is like a boil on the back of the neck or elsewhere that with pricking can heal. More importantly, pricking can reveal the real world dynamics of all that more generally can drive us both in science and religion off track in evolution.

First developed by the remarkable American philosopher Charles Sanders Peirce in the 1860s, grown since to major status in the borderline between philosophy and science, the field of semiotics offers the most highly developed of methods for tracking the relationship of mind space to mind space that follows.[4] I mention this to signal to semioticians the kinship. For the sake of continuity and easier comprehension we will draw our tools for evolutionary surgery from Part II: A Language for Evolution and Revolution.

Let's first put the *field* to work with a picture of the relation of the *personal* mind space of Neo-Darwinian science (**N**) to the *social* mind space of America and the world at large.

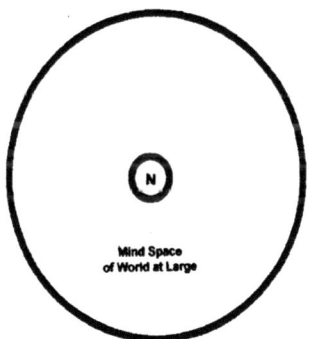

Figure CC 1
Relation of Neo-Darwinian Mind Space to
Mind Space of the World at Large

Here we see how within the total consciousness—that is, the mind space for America and the world at large, indicated by the large circle—the radically limited mind space of NeoDarwinian science was like a tiny island within an ocean of consciousness.[5]

In relation to eugenics, the NeoDarwinian connection was basically the established faith in natural selection acting on random variation as the primary force in evolution. But through the work of Fisher and others we looked at in chapter six had been added the vision of vastly expanding control of the gene pool to improve quality.

If one could breed better pigeons, why not better corn, chickens, indeed everything we eat on a massive, even mechanized scale—all of which came to be, to the benefit of everybody. But if one could do this with such success with everything else, why not solve the problems of crime, politics, poor people, and everything else that troubled us, by breeding better humans?

Now we add a separation of the personal space for Neo-Darwinian science into Innovators (**I**) and Followers (**F**)—with a question beginning to form out in America and the world at large.

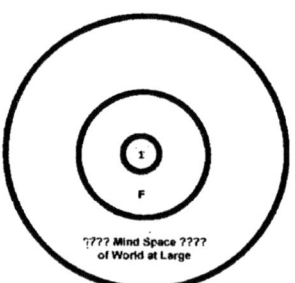

Figure CC 2
Relation of Neo-Darwinian Mind Space to
Mind Space of the World at Large

Here we can begin to see the reduction in discourse, language, constraints, and sense of responsibility that accelerated the eugenics movement and more widely the battle for 20th and 21st century mind. For among the Neo-Darwinian Innovators (**I**, inner core) the exploration of their island in the ocean of mind was advanced through internal debate in an increasingly isolated and private language. Alleles, chromosomes, dominant, hybrid and regressive genes, et cetera, et cetera—the cascade of new words and concepts escalated until by now a biologist must be conversant with thousands of terms at this level of complexity.[6]

But among the Followers—that is, **F**, the next outer core—were and are the scientists in all the other fields, who were and are confronted with the need to display an acceptable familiarity with fundamental concepts and language for something so basic to all of science as evolution.

So what happens?

Here was this incredible array of language we've just barely indicated for biology being loaded in upon scientists already swimming hard to keep from drowning within the complexities of their own escalation of specialist languages. To keep one's head above water, the Followers collapse the complexities of the Innovators into the handy, all-purpose choice of the safely established umbrella term of writ large Natural Selection and Random Variation, and return to their own diverse concerns.

But out there beyond the cloister of theory, within the bewildering and at times horrifying complexity of our present world, were all the questions back then and still now being raised about evolution.

Why Hitlers?

Why do bad people and bad causes prosper?

Why do good people and good causes go under?

What can we do to gain the better world we've been told we're going to reach by switching our loyalty from God to this thing called evolution?

And to get right down to the core of the matter: If we're smart and want to make money, and support our family, and get ahead, to whom or to what does this thing called evolution tell us we should align ourselves? What faith or plan of action must we live by?

As Darwin had lamented, behind the idea of natural selection and variation were too many complexities for unscientific comprehension. Out in the wide world beyond science "natural selection and variation"soon became little more than a phrase to remember for a test, then forget. The world at large did, however, grasp a brutal core meaning. Up came the handy phrase that Herbert Spencer had coined to Darwin's regret. Up came the all-purpose answer for a hard-core economic and political systems leadership. Up came the answer for the model, system, ethos and the leadership for the prevailing D-attractor mindset or paradigm.

Empowering both the eugenics movement and its use by the creationists to lambast science, up came the sacred mantra of mantras: *Survival of the Fittest,* with a snappy late 20th century updating for the ethos of the *selfish gene* and selfishness *uber alles.*

As with the way Einstein's E=mc² became the atom bomb and numerous other examples, we see how theories conceived in the cloistered world of science can become nightmares in a morally ungoverned larger world.

Behind what happened lies the larger question we raised earlier. To whom or to what does this thing called evolution tell us we should align ourselves? What faith or plan of action must we live by?

We are back to chapter ten's key differentiator for evolution. We are back to the question of the degree to which we live within and live by the violence and power for the few of a domination system, animated by the D-attractor, or the peace and plenty for many of a partnership system, animated by the P-attractor.

Indeed, within the evolutionary perspective of millennia, we are looking at something even larger. In the relation of the imbalanced and laggard reality of the science of the 20th century to the visionary potential for science, I think that into the mind of someone a century from now, looking back, may come the image of a dinosaur awakening to the possibility of becoming a bird or a human.

EVOLUTIONARY SYSTEMS SCIENCE, POLITICAL SCIENCE, AND ECONOMICS

In closing this book I realize that for some years now I have been referring to "the new field of evolutionary systems science" without defining it.

The field of systems science originated with biologist Ludwig von Bertalanffy and South African general Jan Smuts early in the 20th century. The vision and purpose was to transcend the customary barriers between all the separate fields of natural and social science, to simply apply whatever findings best fit the question or problem at hand.

Though respected, this approach was generally viewed askance as a rather slippery proposition by all the separate fields, each comfortably locked within themselves by cherished references and separate languages. Over time this new field found a home for itself in its own International Society for Systems Science (ISSS) and a well-financed base in the defense industry in World War II, and afterward in the corporate business world.

This alignment became its primary identity: to tackle big problems with a lot of money involved. But with the arrival of cybernetic, chaos, complexity, and self-organizing process theory, systems science began to take on a much more exciting new identity. Searching for language to describe it, I felt the best term for what was happening was *evolutionary* systems science.

Moving beyond the military, business, and a peripheral academic status among the separate fields of science, here was this new thrust for science linked to the great historic stream of thought exploring reform, revolution, and all other ways of advancing human evolution.

The emerging goal was to go beneath surfaces to discover how to change the present world of injustice and exploitation into something better. The goal was to get at and advance what *should* be, rather than what *did* prevail.

Historic roots, I felt, could be glimpsed in sources as diverse as Jesus, Gautama, Confucius for religion; Heraclitus and Aristotle among the Greek philosophers; *I Ching* for early China, both Marx and Engels and Darwin in the 19th century.

All were fundamentally concerned with evolution, some above all with moral evolution. This to me became the meaning for evolutionary systems science, as the term I coined to provide a semantic umbrella for this very exciting new scientific field.

Evolutionary Political Science and Evolutionary Economics

In this book I'm also, for the first time I know of, naming and beginning to define what I feel has become the vital field of *evolutionary political science*.

I've done this as the logical outcome to a long string of previous books and articles moving toward the conviction expressed in chapter ten.[1] As expressed there in terms of history and advanced evolution theory, I make the case that "we shall fall short, and invite oblivion, until we build the fully human, political, economic, and moral action-oriented theory and story now seeking form within the vision of Darwin's second revolution."

Naming and beginning to define this new field is further needed, however, because of the strong probability that in the twilight of the Super Neo hold on mass mind they will try to do to political science what (as shown in the final chapters of this book) they tried to do to sociology and

153

psychology. That is, I want to pre-empt the highly probable attempt of the Super Neos to slap the word "evolutionary" to "political science,"and in the name of progress again blindly set out to nudge political science backward toward the old paradigm.

What is needed is pointed up by the background to the development of the relatively new field of evolutionary economics. This term was first used by progressive economist Thorstein Veblen.[2] One of the great co-founders of the field of systems science, progressive economist Kenneth Boulding, then launched this new field with papers in journals and books including, in 1981, *Evolutionary Economics*.[3]

In keeping with Darwin's original vision for the completion of theory, both Veblen and Boulding were notably motivated by the moral, human, future oriented, and systems scientific thrust for Darwin's second revolution, rather than by the blinded counter-revolutionary thrust for the Neos and Super Neo at their worst..

THE DARWIN PROJECT

The Darwin Project was formed in 2005 to speed the shift in education and the media from old paradigm Darwinism to the theory and story of evolution of Darwin's long ignored original intention and the thousands of modern scientific studies that corroborate, update, expand, and advance it. In support of this mission a Darwin Project Council was formed of over fifty distinguished American, European, and Asian educators, scientists, and media activists. For concerned people looking for ways to put the past behind us and more speedily gain the world our species deserves and aspires to, see our website: www.thedarwinproject.com.

Darwin Project Council Members

Marcus Anthony
Angeles Arrien
Ralph Abraham
Kenneth Bausch
Richard Bird
Howard Bloom
Raymond Bradley
Alexander
   Christakis
Allan Combs
Gerald Cory, Jr.
Milhaly
   Csikszentmihalyi

Riane Eisler
Duane Elgin
Sally Goerner
Rod Gorney
Thom Hartmann
Hazel Henderson
Mae-Wan Ho
Barbara Marx
   Hubbard
Sohail Inayatullah
Min Jiayin
Jeffrey Kane
Helena Knyazeva

Stanley Krippner
Hans Kung
Ervin Laszlo
Michael Lerner
Daniel Levine
Bill Levis
David Loye
Peter Meyer-Dohm
Ron Miller
Alfonso Montuori
Nel Noddings
Bruce Novak
Karl Pribram

David Loye

Raffi
Robert J. Richards
Ruth Richards
John Robbins
Frank Ryan, M.D
Stanley Salthe

David Scott
Tim Seldin
Christine Sleeter
Joseph Subbiondo
Brian Swimme
Michael Toms

Deceased
Bela A. Banathy
Paul Maclean
John O'Manique

A CHALLENGE FOR BOLD
TEACHERS AND BRIGHT STUDENTS

Would you like to make a historical contribution to science? Progressive philosophy? Or progressive religion?

Strange as it may seem to a teacher with students, or to a student yourself, at the start rather than well along in your intended career, you really can.

This book and others I've written or edited about the new versus the old theory and story of evolution contain descriptions and references to the work of hundreds of scientists and other evolution theorists. They also reveal four things about a remarkable new opportunity for the teacher with a potential team of bright students, the bright student with an excellent teacher and bright prospective team mates on hand, or the teacher with likely faculty team drawn from home base or other schools.

First is how within the new story and theory, like the intermittent sight of the moon on a stormy night, we can glimpse the basis for a guide to a better future. Next is how quickly this tantalizing glimpse is lost in the scatter and contentious tangle of books and studies now both advancing and obscuring our understanding of evolution. Third is how this frustrating situation underlines the urgent need for a 21st century synthesis to update and improve the original neoDarwinian synthesis that set the stage—and too often disastrous effect—for 20th century evolution theory.

Fourth, most striking of all, is the degree to which lack of progress in this direction makes it clear that the new synthesis is *not* likely to come any time soon from presently established theorists and other authorities in science, religion, or philosophy. They're either still fixated in the mangled past or too involved in their personal part of the needed whole. Hence, the opportunity of a lifetime for teachers and students coming relatively fresh to the task.

The 21ˢᵗ Century Synthesis Project

Would you be interested, to echo Star Trek, in going ahead "where no one has gone before?"

What if there could be launched a 21ˢᵗ Century Synthesis Project for teachers and students? With prizes, judging by a group of national and internationally known scientists and educators, publicizing to participants' advantage in building careers, winning submissions published in a book for global distribution?

You've presumably read this book. Or are about to do so. Please think about it. Would you like to explore and help shape such a project?

Take a look at the expanding sketch of prospects on my website and write me.

Website: www.davidloye.com. Email: davidloye@gmail.com.

BUYER'S GUIDE TO *DARWIN'S LOST THEORY*

As frequently noted, a vital data base for the books of this trilogy is *Darwin's Lost Theory*. For the first time *Darwin's Lost Theory* brings together what Darwin actually wrote in page after page of the long ignored "higher order," moral action completion for his theory of evolution. But this essential book is now caught in the Trap of the First Edition, which not only undermined Darwin's *Descent of Man,* but also undercuts full, effective understanding of countless new books today.

As we've seen was the case for *The Descent of Man,* important books are generally, sometimes even radically, improved by their authors in later editions. But for a variety of reasons both mass and elite publication tends to become locked in on first editions, which may well be the worst edition. The result in our time is that the online book sellers market is so flooded with cheap used copies of earlier, lesser editions that the later, better edition can, in effect, drown without a squeak. Why pay $20 for a book you can get for $4.95?

In the case of *Darwin's Lost Theory,* the problem of new versus old is critical. For general readers earlier used copies are fine. But scientists and scholars do not read books in the usual way. A fast scan of the table of contents, then the references, and then—above all in importance—one speeds to the index to spot the page or pages must useful to oneself. Besides the advantage of considerable consolidating, sharpening, updating, striking new cover, and new subtitle Bridge to a Better World, the new third edition of *Darwin's Lost Theory* is the first to have the comprehensive index vital to the scientist, scholar, teacher, and student in all fields.

Determined to see it reach this vital readership and do the good in the world for which it is intended, I've radically cut the price to where, all things considered, it's cheaper to buy *Darwin's Lost Theory* new than used.

THE REST OF THE NEW STORY

The need for mentally, emotionally, morally, and spiritually regrounding our lives at a time of pivotal evolutionary challenge has become inescapably apparent. At the core of this challenge I've found that to untangle and update the story of Darwin and evolution theory has required far more books than I could have imagined at the outset. As of this writing, what over a decade ago began as a single book has expanded into five, with two more to go.

Here are brief synopses.

Darwin's Lost Theory

Contradicting the long embedded stereotype of "survival of the fittest" and "selfish gene" Darwinism, this work is my reconstruction of Darwin's long ignored "fully human, love and moral-action-oriented" completion for his theory of evolution. In page after page of Darwin's own long ignored original writing, Darwin makes the case for the primacy of mutual aid, education, love, and moral sensitivity as higher order drivers for human evolution. The new third edition has been revised and updated with special indexes tailored to the needs of an exceptionally wide range of readers—philosophers, theologians, teachers and students in all fields, as well as scientists and general readers. In Part I: A Young Man's Bold Vision, we meet and get to know Darwin in the critical months during which he first strayed on what became the known theory of evolution, for which he became famous. But of more interest now are all the seemingly contrary insights in his private notebooks, which became the long ignored completion for his theory. In Part II: An Old Man's Surprises, it's 30 years later. We follow Darwin as he writes of how, rather than being slaves of "selfish genes," far more often than we're aware, we're driven by moral sensitivity. Of how, though selfish, we are also driven by love to transcend selfishness. Of how, though fiercely motivated to survive and prevail, we are also driven by a transcendent need to respect and care for the needs of others. Strikingly corroborated by scores of new and earlier modern studies, *Darwin's Lost Theory* shows how recognition of this moral-action-oriented completion for Darwinian theory of human evolution could have changed the 20th century for the better—and can still help us save the 21st.

Darwin in Love

Long scattered throughout Darwin's writings are ninety five of his stories of the love and sex life of an intriguing range of birds and animals of all sizes. They range from the tiny barnacle to a moral band of pelicans, loving pair of orang utans, and a supremely loyal elephant. Particularly fascinating is Darwin's exploration of how the "moral sense" primary to the long ignored completion for his theory of evolution first emerges in the caring rather than the killing of one another far down the evolutionary scale for all of earth's life forms. The charm and delight of Darwin's own love story as a teen-ager and later family man—and the insights and humor of these stories of love and caring behavior, for the first time brought together here—provide an engaging introduction to an inspiring new story of evolution. Written for readers of all ages.

Bankrolling Evolution

Subtitled *A Program for a President, Bankrolling Evolution* explores how applying Darwin's completed theory of love and moral sensitivity to the hard political and economic realities of the 21st century can advance evolution. In terms of the history and psychology of the American Dream, the book contrasts the consequences of the regressive mindset and policies of the Greed Era and G.W.Bush years with progressive hopes for the Obama presidency. My new Global Sounding measure of what advances, checks us in place, or drives us backward in evolution backs up findings with the basic scientific power of measurement.

Measuring Evolution

With subtitle A Leadership Guide to the Health and Weath of Nations, this is a user's guide to the Global Sounding, a new measure of local, national, and global health and wellbeing based on my reconstruction of Darwin's completed theory and the findings of thousands of modern scientists that corroborate it. Historically, this is the first scientific measure of evolution not merely in terms of millions of years in the past, but evolution as we actually know it and must face its huge challenge today. Monitoring fifteen levels and activities for evolution, this new instrument provides a measure of cosmic, chemical, and biological evolution, the evolution of the brain, psychological, sociological, cultural, political, economic, educational, technological, on into moral and spiritual evolution, and the evolution of consciousness and action. This book also includes the new fifteen tenet evolutionary grounded Global Sounding Moral Code and scenarios for use of the new measure and moral code by decision-makers in business, government, politics, science, education, the media, polling, nonprofit and religious organizations, and philanthropists and foundations.

David Loye

Darwin's Second Revolution

Throughout these new books on the "second Darwinian revolution" runs the haunting question of how on earth could the rest of Darwin be lost for over a century? Detailing the dramatic, century-long, battle of progressive scientists against the distortion of Darwin's theory, ***Darwin's Second Revolution*** reveals the astonishing power of the seizure of 20th century mind by the mindset of "survival of the fittest" and "selfish genes" now politically, economically, morally, environmentally, and spiritually driving us toward destruction. This is the first of the three final "Darwin" books for my trilogy Darwin and the Battle for Human Survival. (See www.davidloye.com for covers, table of contents, excerpts, and sample chapters).

Forthcoming during 2011 are these remaining books for the trilogy Darwin and the Battle for Human Survival.

The Battle of the Books

How and why were we denied the Darwinian bridge to a better world for over a century? We go behind scenes within the battle of the books that decade after decade was fought to shape and thereby control the 20th—and now 21st century mind. Out of a probe of 158 books on Darwin and evolution, a scan of the best biographies—including the Desmond and Moore award-winner and Darwin's own autobiography—provides the battleground. Then steadily the battle unfolds between second Darwinian revolutionists and the first turned counter revolutionists. We go behind scenes to look at the armory and impact of books for seventeen champions of the Neo and Super Neo status quo—e.g., E.O.Wilson, Robert Trivers, W.D.Hamilton, Daniel Dennett, Richard Dawkins of the "selfish gene."

We do the same for nineteen of their antagonists, including Julian Huxley, Ashley Montagu, Stephen Jay Gould, David Sloan Wilson. Driving the fateful see saw between evolution and regression, chapter by chapter the drama unfolds of the lockstep battle between evolutionists, the tragic lunacy of the creationists, and the greed for power of corporations and rightwing politicians and religions.

Up Against the Paradigm

As in a spy novel, the book opens with the true story of a secret meeting by scientists during the Cold War, behind the Iron Curtain, in Budapest, to see if something can be done about a potential End Game problem for the 21st century. With global tension exponentially rising, they meet to see if a shift from a "survival of the fittest" to a "mutual aid" theory of evolution might help keep Russia and the United States from mutual nuclear annihilation. Thereafter the true story unfolds of how the scientists and scholars of many fields and many nations have worked to update and expand the theory and story of evolution to meet the enormous challenge of the 21st century. We see how meeting in conferences with others throughout Europe, North America and Asia, interfacing via the then new internet, bit by bit they gradually expand the data base for a theory of evolution pointed toward peace and plenty rather than more of what devastated the 20th century environmentally, socially, economically, politically, morally, and spiritually. In this book we meet the visionaries, innovators, and experts who, in going up against the paradigm, laid the groundwork and defined the challenge for 21st century science and society.

ABOUT THE AUTHOR

Spanning much of the 20th into the 21st century, mine has been a life of creativity in pursuit of paradigm-puncturing scientific and social advance.

For the past twenty years I've been globally involved with other scientists in developing the cutting edge fields of chaos, complexity, and integral theory from the powerful new perspective of evolutionary systems science. I am a co-founder of two international organizations for the expansion of evolution studies: The General Evolution Research Group, and The Society for Chaos Theory in Psychology and the Life Sciences; co-founder with my partner—pioneering cultural evolution theorist and well-known author of *The Chalice and the Blade*, Riane Eisler—of The Center for Partnership Studies (www.partnershipway.org); founder of The Darwin Project (www.thedarwinproject.com), with a Council of more than 50 progressive American, European, and Asian scientists, educators, and media activists—and now, as of this writing, soon to become founder of the Moral Transformation Project, with new website (www.moraltransformation.com) currently under construction.

Darwin's Lost Theory

In the 1980s I began to uncover and write of what has been the chief notable item for my work to date: Darwin's long ignored moral- and action-oriented completion for his theory of evolution.

With widening corroboration by biologists, brain, social and systems scientists, my reconstruction of the "rest of Darwin" was hailed as a major contribution to our understanding of evolution (See What the Experts Say, www.davidloye.com). Even so, my basic grounding book—head on challenging the disastrously entrenched paradigm of "survival of the fittest" and "selfish gene"—was year after year steadily rejected by publishers during the late 20th century Greed Era for America. Darwin's "lost theory" of love and moral sensitivity as prime drivers for human evolution first gained formal book publication in Germany and China. Then late in 2004, the State University Press of New York (SUNY Press), published *The Great Adventure: Toward a Fully Human Theory of Evolution.*

A New Faith and World View

"In times like these a new worldview often arises at the margins of power, at the periphery of the action unfolding on the main stage," internationally eminent psychologist Mihaly Csikszentmihalyi writes in the foreword to this book by myself with chapters by eleven other members of the General Evolution Research Group and The Darwin Project Council.

"The themes introduced by the authors are likely to be among the central ones of any new world-view," Csikszentmihalyi wrote. "David Loye's central insight, which motivates this book, is in my opinion right on the money. The organizing principle of the new faith—a faith of human beings about human beings—is evolution itself. Not the traditionally taught evolutionary scenario dominated by competition and selfishness, but an understanding closer to the original Darwinian one that sees cooperation and transcendence of the self as the most exciting parts of the story."

Thereafter, at last came publication of *Darwin's Lost Theory; Darwin on Love* (with updated and retitled new edition *Darwin in Love* scheduled for publication on Valentine's Day, 2011); *Bankrolling Evolution;* and *Measuring Evolution.* Now underway is publication of the new trilogy Darwin and the

Battle for Human Survival (*Darwin's Second Revolution, The Battle of the Books,* and *Up Against the Paradigm;* see www.davidloye.com, for covers, table of contents, index, and excerpts).

A Life Shaped by Our Times

Behind these books—and the dreadful need we face to promote ourselves if in any way involved in "going up against the paradigm"these days—lies the life that brought me to this place at 85.

I can see a concern for others was ground in by my childhood during the Great Depression. While a very young news correspondent with the U.S.Navy in the closing years of World War II, I docked and roamed the same ports in South America that Charles Darwin visited one hundred years earlier on the famous voyage of the Beagle. I became a television newsman during the Edward R. Murrow days. I wrote the national award-winning *The Healing of a Nation* and gained my doctorate in psychology in early middle age. While a Princeton and UCLA School of Medicine faculty member, I was the research director for major studies of political values, the use of the brain and mind in prediction, and the impact of movies and television on adults.

I first gained a foothold in the world of action science as a futurist. My earlier books include *The Leadership Passion, The Knowable Future, The Sphinx and the Rainbow,* and *An Arrow through Chaos.* Out of this phase came an increasing involvement with systems science, chaos and complexity theory, and, as editor, *The Evolutionary Outrider: The Impact of the Human Agent on Evolution.*

Now midway into my eighties, besides and beyond the "Darwin books," I'm writing and publishing two other "cycles." One set of books, already published, explore what most of us would agree is the best of life: travel, history, mystery, humor, family life, and love. So far, this set includes *Brave Laughter, Return to Amalfi, The Parable of the Three Villages,* and *Tangled Tales of the Book Trade, or the Mystery of the Missing Century,* with a children's book, *Grandfather's Garden,* in the works. Of increasing popularity is *3,000 Years of Love: The Life of Riane Eisler and David Loye,* my joint biography of the early lives and the adventure of life together in science, social action, and love for my wife Riane Eisler and myself. In *100 Days of Love* and *1001 Days of Love,* I've also published two books of love poems—so far 158 in number, already four more than for the benchmark 154 for Shakespeare's famous sonnets!

The other set of books is of what I originally set out to mainly do. Diverted for over a decade by the fascination but also drain on mind, energy, and finances of the Darwin books, I will return to the Moral Transformation Project. Currently under construction is the website www.moraltransformation.com.

To be displayed online in an innovative and interactive new way are *The Parable of the Three Villages* and *The River and the Star: The Lost Story of the Great Explorers of the Better World,* both already in print and available worldwide through book stores and online book sellers. To this, as an online work in progress, for viewing as can fascinate us in watching a building under construction, I will add *The Glacier and the Flame I: Rediscovering Goodness; The Glacier and the Flame II: Redefining Evil; The Glacier and the Flame III: A Fragment of a Vision; The Science of Evil;* and *Moral Sensitizing: A guide to a new method of learning and therapy for teachers, counselors, ministers, and self-healers.*

For more of my scientific, educational, and personal life, see biography, www.davidloye.com

NOTES

Prologue

1. See www.davidloye.com.

2. "I have been led to put together my notes, so as to see how far the general conclusions arrived at in my former works were applicable to man," he wrote on the *second page* of *The Descent of Man*.

3. Ibid, pp 101,102.

4. Ibid, p.531.

5. Eric Hobsbawm, War and Peace, *The Guardian,* February 23, 2002.

6. Additional to the evidence marshaled in this book and trilogy, behind this picture of an alignment to regression for Republicans and an alignment to progression for Democrats is the result of my earlier study of the pathology of the G.W.Bush years and policies in *Bankrolling Evolution*, and development of the Global Sounding measure of global health and wellbeing, and the Global Sounding Moral Code, outlined in *Measuring Evolution.*

Chapter One

1. Romanes, *Darwin and After Darwin,* Vol.1, pages 9-10.

2. Ibid, page 10

3. Ibid, 11

4. Ibid

5. Ibid, 12

6. Ibid, 11-12

7. Richards, *Darwin and the Emergence of Evolutionary Theories of Mind and Behavior*, 336.

8. Ibid

9. Ibid

10. Ibid

11. Ibid

12. Darwin, *Origin of Species*, 239.

13. Ibid

Chapter Two

1. See chapters two, 11, 15, and 17 in *Darwin's Lost Theory*.

2. In philosophy this was the assertion of some of the great turning-point thinkers of both the French and the Scottish Enlightenment, which in the next generation became the watershed moral statement of Immanuel Kant. In modern science this affirmation has been massively confirmed, for example by the research of one of the 20th century's greatest brain scientists, Paul MacLean, and others I write of in the other books of this venture.

3. See Smithsonian National Museum of Natural History: http://paleobiology.si.edu, and references for MacLean, Karl Pribram, Joaquin Fuster, Ward Halstead.

4. Evolutionary systems science is a comparatively new field emerging from the earlier established field of systems science driven by the thrust of the late

20th century explosion of cybernetic, chaos, complexity, and self-organizing theories. See Evolutionary Systems Science, Political Science, and Economics in Reflections ending this book.

5. See Margulis and Sagan, *The Mystery Dance.*

6. See MacLean, *The Triune Brain in Evolution.*

7. See Swimme and Berry, *The Universe Story.* This book has an excellent time line for all periods of evolution.

8. Ibid.

9. MacLean, *The Triune Brain.*

10. See Durant and Durant, *The Story of Civilization;* Toynbee, *A Study of History;* Tarnas, *The Passion of the Western Mind.* While moral evolution was not their primary focus, in books such as these one can easily track the refinement of ethics and morality through the build up of habit century by century over the past 5,000 years. (Before then, see Eisler, *The Chalice and the Blade,* for a different story).

11. Here one could rightfully cite practically the entire field of psychology comprising some millions of books by now. E.g., Mednick for learning, Lewin and Brown for social psychology, Erikson for development.

12. See Maslow, *Toward a Psychology of Being;* Schneider, Pierson, and Bugental, *Handbook of Humanistic Psychology;* Wilber, *Integral Psychology.*

13. Darwin, *Descent (2nd edition),* 39. For elaboration, see chapter 25, *Darwin's Lost Theory.*

14. See Jantsch, *The Self-Organizing Universe;* Lorenz, Irregularity; Prigogine and Stengers, *Order Out of Chaos.*

15. This was the historical and religious significance of Moses and the exodus of the enslaved Jews from Egypt, later the emergence of Jesus. It was the scholarly activist assertion of Voltaire and other French Enlightenment thinkers

that emboldened Washington, Jefferson, Franklin and the other founders of America. It thundered out of Marx and Engels in launching the science of activism. It quietly emerged as "functional autonomy" in the psychology of Gordon Allport. Then out of the spread of chaos theory it exploded, as "autocatalysis" in the work of Belgian thermodynamicist Ilya Prigogine, "autogenesis" for Hungarian biologist Vilmos Csanyi and computer scientist Gyorgy Kampis, and "autopoesis" in the work of Chilean biologists Humberto Maturana and Francisco Varela.

16. See chapters 22, 23, in *Darwin's Lost Theory.*

17.See *Darwin's Lost Theory* for analysis and rationale.

18. Desmond and Moore, *Darwin's Sacred Cause.*

19. Darwin, *Descent 2ⁿᵈ edition,* 107. See p. 5, endnote 10, in Prologue for Loye, *Darwin's Lost Theory,* for elaboration. Wholly contrary to the view of both the Creationists and Neo and Super-Neo Darwinian science, Darwin actually deserves a special place in the long line of highly practical visionaries linking morality to spirituality. For within the context of many centuries of history and moral philosophy he looms as the first to provide a scientific grounding for the line from Jesus and Gautama to the Dalai Lama and the great German theologian Hans Kung, whose Global Ethic is a compelling update for our time.

Chapter Three

1. See Clark, *The Survival of Charles Dawin,* 235-236.

2. Romanes, *Darwin and After Darwin,* vol.2, 12.

3. Wallace in a letter to Charles Darwin April 18, 1869: see Richards, *Darwin and the Emergence of Evolutionary Theories of Mind and Behavior,* 244.

4. See Richards, *Darwin and the Emergence of Evolutionary Theories of Mind and Behavior,* 174-186, for Wallace on spiritualism. Also Smith and Beccaoni, *Natural Selection and Beyond: The Intellectual Legacy of Alfred Russel Wallace.*

5. Ibid, 179, for complete list—including Darwin!

6. Wallace, *A Defense of Modern Spiritualism*, 56.

7. *The Daemon* by Anthony Peake is an interesting exploration of this contention by Wallace.

Chapter Four

1. See Richards, *Darwin...*, chapter eight, for an excellent account of Morgan's life and works.

2. Morgan's Canon, *International Congress of Experimental Psychology,* second session.

3. See Richards, *Darwin...*, chapter ten, for a dramatic developmental account of The Baldwin Effect. Also. more recently, Weber and Depew, *Evolution and Learning: The Baldwin Effect Reconsidered.*

4.As for Wallace, on hearing of what for a time was heralded as the exciting new Morgan-Baldwin-Osborn idea, Wallace proclaimed that "all the theoretical objections to the 'adequacy of natural selection' have been theoretically answered."

5. Richards, Darwin..., 402, even including a footnote acknowledging a similarity to earlier work by Morgan.

6. Romanes, *Darwin and After Darwin,* Vol.2, 153.

7. Ethel Romanes, *Life and Letters of George John Romanes,* 327-341.

8. James, *Diary*, entry for 30 April 1870.

9. James, *Principles of Psychology*, vol 1, 141.

10. Richards, *Darwin*, 422.

11. James, *Bain and Renouvier*, 369.

12. James also explored the difference between the brutal first half and the second transcendent half for Darwin's completed theory in another way generally forgotten these days. This was his early observation of the division in all matters of contention into the *tough-minded* versus the *tender-minded,* explored by scores of psychologists in the mid-20th century.

13. See chapter eight, in which we'll explore the entry of chaos theory into the development of evolution theory.

14. The lurid detail is covered in Richards' excellent account in *Darwin...*, pp.495-503.

15. See Richards, *Darwin,* 495-503.

Chapter Five

1. This would place the time of writing for Kropotkin in the late teens or early 20s.

2. Kropotkin, *Ethics,* 12.

3. Ibid, 13.

4. Ibid, 13-14.

5. Ibid, 16.

6. Durant, Greek book

7. See Durant, *The Age of Greece;* Freud, *Civilization and its Discontents;* Cannon, *The Wisdom of the Body.*

8. See chapter 10, Liberals and Conservatives vs. Regressives, in Loye, *Bankrolling Evolution.* An excerpt: "As I explored in *The Leadership Passion,* in terms of the dynamics of human evolution, progressives or liberals act as the *norm-changers,* who advance evolution by decrying the gap between whatever

presently exists and the ideal they can see is both desirable as well as distinctly within human capability ... By contrast, the "true," bona fide, morally oriented and sensibly functional conservatives act as *norm-maintainers*. Leery of change, fearing that it can make things worse rather than better, they dig in to keep in place whatever, however imperfect it may be, presently exists. In this regard, they serve an evolutionary function in preserving the investment of thousands of years of evolution for our species to get this far." I note that "new ways must ... be found to innoculate conservatives against invasion by the virus of regression ... to whoever or whatever promises order ... many conservatives will automatically pledge their loyalty and their money. And thereby the pathology of leaders, enablers and followers gains up to the point of the overwhelming expose—as in the case of U.S.President Nixon and the 'Nixonians,' of whom scores were returned to power in government during the Bush years." pp.98-99.

9. Kropotkin, *Ethics,* 13.

10. Ibid, 41.

11. Ibid, 47-48.

Chapter Six

1. Clark, *The Survival of Charles Darwin.* See chapter eleven.

2. Ibid.

3. Eldredge and Gould, Punctuated Equilibria.

4. Ibid. See chapters seven and eleven.

5. Ibid, 252.

6. Clark, *The Survival of Charles Darwin*, 257-258.

7. Ibid, see chapter twelve.

8. Ibid, 257-258.

9. Ibid, 279-280. Regarding the "war of all against all," this was Darwin's long ignored and radically modifying key observation in *Origin of Species:* " ... the most important of all causes of organic change is one which is almost independent of altered and perhaps suddenly altered physical conditions, namely, the mutual relation of organism to organism,—the improvement of one organism entailing the improvement or the extermination of others ...," 372-373.

10. Regarding "unoccupied, or poorly occupied places in nature," this was Darwin's observation at around age 28 in his early notebooks: "One may say there is a force like a hundred thousand wedges trying force every kind of adapted structure into the gaps in the oeconomy of nature. or rather forming gaps by thrusting out weaker ones." See Gruber and Barrett, *Darwin on Man,* p.456.

11. Clark, *The Survival of Charles Darwin*, chapter twelve.

12. Ibid, 290.

13. As Stephen Jay Gould hammered away at in *Dinosauer in a Haystack* and *The Structure of Evolutionary Theory,* in the last five chapters of his key book *The Genetical Theory of Natural Selection* Fisher concluded with a "single coherent (if fatally flawed) argument in eugenics."

14. See Hofstader, *Social Darwinism in American Thought.*

15. Huxley, *Evolution.*

16. Clark, *The Survival of Charles Darwin*, 323.

17. See Kevles, *In the Name of Eugenics;* Black, *The War Against the Weak.*

Chapter Seven

1. I remember writing it, but cannot find it! Free book to the first reader who finds this pungent observation in one of my books.

2. Gilgoff, D. Gallup 'Darwin's Birthday' Poll: Fewer than Four in Ten Believe in Evolution.

3. One might raise the question what about William James, John Dewey, and in particular Jean Piaget during those years—but none were formal Neo-Darwinians.

4. This trend can be seen in the work, e.g., of George Lakoff, Michael Lerner, and Riane Eisler. I've set out to give it full scale treatment in the development of moral transformation theory in forthcoming trilogy *The Glacier and the Flame: I, Rediscovering Goodness; II, Redefining Evil; III, A Fragment of a Vision.*

Chapter Eight

1. For a sample of the incredible number of books generated by Laszlo, see Laszlo, *Evolution: The General Theory; The Choice;* and *The Akashic Field.* See Loye, Editor, *The Evolutionary Outrider: The Impact of the Human Agent on Evolution, Essays Honoring Ervin Laszlo,* for brief Laszlo biography. As of this writing, Laszlo has completed a beautifully written autobiography sure to be published at some point within 2011.

2. See A Brief History of the General Evolution Research Group in Loye, *The Great Adventure: Toward a Fully Human Theory of Evolution,* or in www.thedarwinproject.com.

3. Ibid.

4. Underlying its mirroring surface in science was and is the power of a new language wedding the most sophisticated understanding of mathematics, as well as powerful imagery and metaphors, to the science and understanding of evolution at all levels, ascending from the atom and the gene of natural science to the beliefs and ideologies of social science, further embracing the moral quest of religion and the life quest of the humanities. It was and is the power of this new language radically advancing the power of systems science to transcend the old barriers splitting science into the sprawl of tight little feudal baronies separate from and often hostile to each other.

5. Beginning with a short stint as a shipboard Navy news correspondent in WWII, I went on to become a television newsman in the Edward R.Murrow days, unsuccessfully writing novels, poems, and screenplays on the side. See Bio, www.davidloye.com.

6. Psychologist Mary Henle, my favorite mentor at the New School for Social Research.

7. See Marrow, *The Practical Theorist,* and Loye, *The Healing of a Nation.*

8. A particularly important aspect of Lewin's field theory for our purposes was his focus on forces that either drive movement toward a goal (helping forces) or block movement toward a goal (hindering or hurting forces). Inspired by the language that had liberated the science of physics (for example, vectors, trajectory, phase space), Lewin developed a new verbal and visual language for an action-oriented social science. Unfortunately this was before the computer revolution. Had he and his students had the speed and graphic power of the computer at hand, they would have transformed both social science and sizeable chunks of society.

9. To couple the visual with a mathematical language, Lewin developed his famous formula $B = f(P, E)$—that is, Behavior is (=) a function (f) of Person (P) and Environment (E).

10. The by now classic capturing of this fact of modern life was futurist Alvin Toffler's *Future Shock.*

11. Lorenz, Irregularity: A Fundamental Property of the Atmosphere.

12. After Lewin's death, through his students this new approach for social and systems science spread into the language of flow charts in business and other fields. My thoughts here come out of my current work aimed at applying Lewin's work to the development of a new visual language for evolution.

13.As noted earlier, this was an expansion for theory particularly meaningful to me because of a small part I played in providing a pioneering bridge between natural and social science. See Loye and Eisler, Chaos and Transformation: The Implications of Natural Scientific Nonequilibrium Theory for Social Science and Society.

180

14. See Abraham and Abraham, *A Visual Introduction to Dynamical Systems Theory for Psychology,* II-80.

15. In this example, first we see the kind of "portraits" and names their discoverers gave to basic evolutionary configurations for everything from the evolution of emotions, thoughts, and actions to the formation of clouds, the patterns for storms, and quite likely everything we know as the history of life on this planet. In the corresponding differences for the "time series" we see how chaos theorists track critical paths for how these configurations take shape over time. Then last, in the "spectra," we see a measure for speculating on potential power of impact.

16. How fascinating life is in retrospect! In my early professional life, beginning with *The Leadership Passion: A Psychology of Ideology*, I became involved in both the study of how we predict the future and the invention of new methods of forecasting. *The Knowable Future, The Sphinx and the Rainbow, An Arrow Through Chaos,* many articles in journals, and *Making It in the Dream Factory* (scheduled for publication in 2012) followed. Hence, for me chaos and complexity theory were like tickets through the rabbit hole into wonderland.

17. See Schlesinger, Tides of American Politics, in *Paths to the Present.*

18. Toward this end I have developed the Global Sounding measure of global health and wellbeing and the Global Sounding Moral Code (see *Measuring Evolution, Bankrolling Evolution,* or www.davidloye.com, or forthcoming www.moraltransformation.com). In the early formation of chaos theory much was made of the idea that the operation of the strange attractor showed that chaos theory revealed it was impossible to predict the future. In *An Arrow Through Chaos*, and in Prediction in Chaotic Social, Economic, and Political Conditions, and How Predictable is the Future, I've shown at great length the fallacy of this notion. Using the disaster of the G.W.Bush presidency as a test case, *Bankrolling Evolution* and *Measuring Evolution* demonstrate the predictive power of the Global Sounding measure of basic indicators on fifteen levels for evolution, ranging from cosmic and biological to social, political, and moral and spiritual evolution.

Chapter Nine

1. This is a selection and reduction of statistics on the devastation of World War II from a comprehensive article in Wikipedia: http://en.wikipedia.org/wiki/World_War_II_casualties.

2. See Loye, *3,000 Years of Love.*

3. See Jay, *The Dialectical Imagination;* Adorno, Frenkel-Brunswick, Levinson, and Stanford, *The Authoritarian Personaltiy;* Loye, *The River and the Star, The Leadership Passion.*

4. Loye, *The Evolutionary Outrider, The Great Adventure, 3,000 Years of Love, Up Against the Paradigm.*

5. Eisler, *The Chalice and the Blade, Sacred Pleasure, Tomorrow's Children, The Power of Partnership, The Real Wealth of Nations.*

6. See Loye, *The River and the Star, The Science of Evil, The Glacier and the Flame* trilogy.

7. Adorno, *et al, The Authoritarian Personality.*

8. Benedict, The Synergy Lectures.

9. Eisler, *The Chalice and the Blade, Sacred Pleasure, Tomorrow's Children, The Power of Partnership, The Real Wealth of Nations.*

Chapter Ten

1. See Hitler, *Mein Kampf.* For Goebbels, Richard Evans, *The Third Reich.*

2. See Kevles, *In the Name of Eugenics;* Black, *War Against the Weaker;* Gould, *The Flamingo's Smile, Dinosaur in a Haystack.*

3. Semiotic? See The Controversial Connection in ending Reflections for definition.

4. Henley, *Invictus.*

5. Level by ascending level for evolution, the contrast between goals for the D-attractor versus P-attractor human being and planetary future can be seen in the Global Sounding measure and the Global Sounding Moral Code (see Loye, *Measuring Evolution, Bankrolling Evolution,* www.davidloye.com.

6. Does this really belong in evolution theory? In terms of political parties, is this comparison in any way biased, unfair, or unscientific? For case studies of the certifiable pathology that engulfed the Republican leadership and party, see Blumenthal, *Republican Gomorrah.* For a wider and deeper probe, see The Pathology of Leadership, The Pathology of Enablers and Followers, The Pathology of Regressive Money, and The Pathology of Consequences in Loye, *Bankrolling Evolution.*

7. See Maslow, *The Psychology of Science;* Myrdal, *Objectivity in Social Research.*

8. As of this writing, for nearly forty years I have been groping to ground the relevance of politics to human evolution in *The Healing of a Nation, The Leadership Passion, The Knowable Future, The Evolutionary Outrider,* and most recently in the development of the Global Sounding integrated measure of fifteen levels and activities for evolution, and the Global Sounding Moral Code explored in *Measuring Evolution* and *Bankrolling Evolution.* Others I find moving in the same direction include cultural transformation theorist Riane Eisler, in *Sacred Pleasure: Sex, Myth, and the Politics of the Body* and *The Real Wealth of Nations;* George Lakoff in *Moral Politics* and *The Political Mind*; Michael Lerner in *The Politics of Meaning* and *The Left Hand of God;* David Prindle in *Stephen Jay Gould and the Politics of Evolution.*

9. See Loye, *The Great Adventure: Toward a Fully Human Theory of Evolution,* with foreword by Mihaly Csikszentmihalyi, and chapters by Ervin Laszlo, Stanley Salthe, Raymond Bradley, Riane Eisler, Sally Goerner, Ken Bausch, Aleco Christakis, Alfonso Montuori, Allan Combs, and Ruth Richards.

Chapter Eleven

1. Wilson, *Sociobiology.*

2. Wilson wrote that humans always have been characterized by "aggressive dominance systems, with males generally dominant over females." He argued further, "Even with identical education and equal access to all professions, men are likely to continue to play a disproportionate role in political life, business and science." Such statements prompted a firestorm of protests from feminists and humanists, and some critics saw an ethnocentric or racist basis to his judgments about the determination of behavior.

3. Ibid, 562.

4. Loye, *The River and the Star: The Lost Story of the Great Explorers of the Better World.*

5. MacLean, *The Triune Brain in Evolution.* See also Loye, *Darwin's Lost Theory;* The Moral Brain; Charles Darwin, Paul MacLean, and the Lost Origins of 'The Moral Sense': Some Implications for General Evolution Theory.

6. Darwin, *Descent* (second edition), 48.

7. By 2009, 82 percent of Americans were Christian according to the Association of Religious Data Archives; the total population for the United States was 307 million according to the U.S.Census Bureau; hence 250, 340,000 for an estimated number of Christians in America.

8. Trivers, *Social Evolution,* 456.

9. Ibid, 457.

10. Ibid.

11. Darwin, *Descent* (second edition), 101.

12.It had further stumped Neo-Darwinian icons J.B.S.Haldane, Ronald Fisher, and Sewell Wright in the drive to unite Darwinian Natural Selection with Mendelian genetics we looked at in chapter six.

13. Hamilton, The Evolution of Altruistic Behavior.

14. Darwin, *Descent* (second edition), 90: "We are thus impelled to relieve the sufferings of another, in order that our own painful feelings may be at the same time relieved ... But I cannot see how this view explains the fact that sympathy is excited, in an immeasurably stronger degree, by a beloved, than by an indifferent person."

15. Dawkins, *The Selfish Gene*.

16. Weiner, *Cybernetics: Or the Control and Communication in the Animal and the Machine.*

17. See Myrdal, *An American Dilemma,* for first definition of vicious cycle or circle. For virtuous circle, the first time I encountered this vital concept was in the early writings of Hazel Henderson, hence I would tend to credit her for the "first" here.

18. Dawkins, *The Blind Watchmaker.*

19. I introduce and define *systematic transvaluation* as a change of meaning for a concept as a matter of routine inbuilt strategy or tactic for control of mind by an over-riding paradigm or system. This would have two variants: a degrading D-systems transvaluation and an inspiring P-systems transvaluation.

20. Dawkins, *The Selfish Gene,* 139.

21. Ibid, 2.

22. Ghiselin, *The Economy of Nature and the Evolution of Sex*, 247.

23. Monod, *Chance and Necessity,* 180.

24. Ibid.

Chapter Twelve

1. Huxley, *Essays of a Humanist*, 34.

2. Ibid.

3. Huxley, *Evolutionary Ethics.*

4. Huxley, *Touchstone for Ethics.*

5. Huxley, *Essays of a Humanist.*

6. Dobzhansky, Cultural Evolution, in David L.Sills, Ed., *International Encyclopedia of the Social Sciences*, 236.

7. In chapters nine, ten, and eleven, we saw how through systematic transvaluation, from the isolated mind space of science to the hurley-burley of politics and economics, the cumulating power of the D-attractor earlier helped lead to Hitler and World War II. By applying the same approach to the re-emergent D-attractor underlying the increasingly radical backward thrust in late 20th century America, we can see why what has been generally written off as no more than the "conservative shift" from the 1970s on into the early 21st century led considerably beyond mere conservatism to the regressive disaster of the G.W.Bush years.

8. Dobzhansky, Nothing Makes Sense in Biology Except in the Light of Evolution, 125-129.

9. Salthe, *Development and Evolution.*

10. See Prindle, *Stephen Jay Gould and thePolitics of Evolution.*

11. Montagu, *The Nature of Human Agression*

12. Gould, *Ever Since Darwin.*

13. Ibid, 258.

14. Ibid, 271.

15. Lewontin, Rose, and Kamin, L. *Not in Our Genes,* 273.

16. Ibid, 290.

17. Gruber and Barrett, *Darwin on Man.*

18. See details in Book II: *The Battle of the Books.*

19. Richards, *Darwin and the Emergence of Evolutionary Theories of Mind and Behavior.*

20. Ibid, 207-208.

21. Ibid, 218.

22. Kauffman, *At Home in the Universe.*

23. A year earlier, similarly keying to the challenge posed by Jacques Monod and the frightful idea of being left "alone in the universe," Princeton physicist John Wheeler came out with a book with the same title.
Wheeler's book was about the older work that comes to life in retroflection—the world of the great physicists of an early time, with whom he had worked as himself considered a major contributor among them.

24. See Loye, Brief History of The General Evolution Research Group, in *The Evolutionary Outrider, The Great Adventure,* and the website for The Darwin Project, www.thedarwinproject.com.

25. Jantsch, *The Self-Organizing Universe.*

26. Capra, *Uncommon Wisdom.*

27. Capra, *The Web of Life.*

28. I have tried unsuccessfully to identify the source for this excellent quote other than Kauffman's impression it went on to be rewritten into his book *Reinventing the Sacred.*

Chapter Thirteen

1. In this regard, the classic case was the response of U.S. President G.W.Bush to the terrorist attack that brought down the twin towers of the Wold Trade

Center in New York City, collapsed thereafter into the code word *9/11,* to commemorate the date: September 11. Earlier President Kennedy had seized the hearts and mind of America and the world with his famous "Ask not what the country can do for you, but what you can do for your country." In sharp contrast for both person and era, President Bush called upon America to go shopping. "Get down to Disney World in Florida," he urged just over two weeks after 9/11. "Take your families and enjoy life, the way we want it to be enjoyed." *Washington Post,* October 5, 2008; Page B03.

2. See note 8, chapter nine.

3. As we'll see in Book II, for philosophic back up they picked the Hobbes and Nietzsche of the war of all against all, rather than—as Darwin keyed to—the Kant, Hume, Hutcheson, and Jesus, Gautama, and Mencius of the inbuilt moral sense.

4. In the *Huffington Post* in 2010 biologist Michael Zimmerman quotes this statement by Richard Dawkins: "No self-respecting person would want to live in a society that operates according to Darwinian laws. I am a passionate Darwinist, when it involves explaining the development of life. However, I am a passionate anti-Darwinist when it involves the kind of society in which we want to live. A Darwinian state would be a Fascist state."

5. Lewin, *Field Theory in Social Science.* Also Loye, *The Healing of a Nation.*

6. Barkow, Tooby, and Cosmides, *The Adapted Mind,* 52.

7. This, for example, is a litany of some of the names and works that every adequately and decently trained psychologist of learning in the 20th century was expected to easily carry in her or his head.

8. Barkow, Tooby, and Cosmides. *The Adapted Mind,* 123.

9. Ibid.

10.Here, I would suggest, can be glimpsed not only the colossal arrogance that alienated a majority of social scientists from the eruption of evolutionary psychology, but more importantly, a warning of what came to be. For one doesn't have to go far to find what soon thereafter became the vicious cycle of teach to the test for American education. I submit that in this passage we are looking at an instance of the systematic transvaluation whereby out of the ostensibly amoral mind space of science, comes blind support for the corruption of the political, economic and religious mind space, which in turn becomes a tool for the twilight attempt at human engineering during America's astonishing entry into the 21st century.

Epilogue

1. Darwin, *Origin of Species,* 243.

2. See Two Stream versus Single Stream Theory *and Consequences* in End Documents for Book II.

Reflections and An Invitation

Romanes on Darwin's Legacy

1. Romanes, *Darwin and After Darwin,* vol.2, p.1.

2. Romanes, *Darwin and After Darwin,* vol.1, 270-271.

3. Ibid.

4. Romanes, *Darwin and After Darwin,* vol.2, 30.

5. Ibid.

6. Ibid, 269.

7. Ibid.

8. Ibid.

9. Vol.1, ll.

10. Ibid, 384.

11. Ibid, 282.

12. Ibid, 418.

13. An interesting example is the work of the progressive theologian Hans Kung, whose book *The Beginning of All Things: Science and Religion* explores the positive relation between science and religion. Another is the work of the progressive evangelist Michael Dowd, who travels the country with the message of his book *Thank God for Evolution.* Perhaps most impressive, however, was the creation of the Clergy Project Letter by biologist Michael Zimmerman, a statement of support for a new working partnership between progressive science and progressive religion, by 2010 involving global membership of 12,000 ministers and 850 scientists.

14. Romanes, *Darwin and After Darwin,* Vol.1, 418.

15. Ibid, 357-358.

The Battle for Heart and Much Larger Mind

1. Jeans, *The Mysterious Universe,* 137.

2. Wilber, *The Atman Project, Up From Eden.*

3. Laszlo, *The Akashic Field, The Akashic Field Experience.*

4. de Quincey, *Radical Nature.*

5. Lerner, *The Left Hand of God.*

6. Loye, *The Evolutionary Outrider: The Impact of the Human Agent on Evolution,* with chapters by Ervin Laszlo, Fritjof Capra, Mae-Wan Ho, Karl Pribram, Alfonso Montuori, Maria Sagi, Raymond Trevor Bradley, Mauro Ceruti and Telmo Pievani, Riane Eisler, Ralph Abraham, Hazel Henderson, Paul Ray, and myself.

7.Working in tandem, joining the known first half with the lost but now regained second half for Darwin's vision, the world we seek becomes possible through partnership between natural and social science, and between progressive science and progressive religion.

8. Henderson, Houston, and Hubbard, *The Power of Yin.*

9. Ibid, 174.

10. Ibid, 175.

11. Ibid, 12.

12. For Eisler, the radically differing structures for domination and partnership systems depend on how gender roles and relations are socially constructed. Rejecting the conventional idea of "matriarchy" versus "patriarchy"as merely the two sides of the same domination coin, she developed the neologisms "gylany" and "gylanic," from the Greek *gyne* for woman and *andros* for man, to semantically link the two in partnership. Likewise, "androcracy" and "androcratic," or *andros* for the man alone ascendent model for domination systems, becomes the replacement for "patriarchy."

13. Eisler, *The Chalice and the Blade,* 203. For the joint biography of Riane and myself, see Loye, *3,000 Years of Love.*

14. Kung and Kuschel, *A Global Ethic.*

15. Noddings, *Caring: A Feminine Approach to Ethics and Moral Education.*

16. Noddings, quote from Nel Noddings, Power Point Presentation for her Educational Psychlogy classes, year unknown.

17. Wilber, *The Spectrum of Consciousness.*

18. Wilber, *A Brief History of Everything*, 42-43.

19. Krippner, in Laszlo, Editor, *The Akashic Experience.*

20. Radin, quoted by Larry Dossey in *The Akashic Experience*, 234.

21. Robbins, The Food Revolution, on website: foodrevolution.org.

22. Caldicott, *Nuclear Madness*.

23. Obama, *Dreams from My Father*, 429-430.

24. Hertsgaard, Who Killed Petra Kelly?

The Controversial Connection

1. See Kennedy, *Darwins Deadly Legacy*. Weikart, *From Darwin to Hitler*.

2. See Forrest, *Understanding the Intelligent Design Creationist Movement*. Zimmerman, Combating the Fifth Wave of Creationism. Zimmerman and Loye, The New Wave of Creationism.

3. For Gould's track of the link between survival of the fittest Darwinism and eugenics, see brief development in *Ontology and Phylogeny,* Carrie Beck's Daughter in *The Flamingo's Smile;* The Smoking Gun of Eugenics, in *Natural History* and *Dinosaur in a Haystack;* and *The Structure of Evolution Theory*. For the horrendous story of the devastation of the eugenics movement in America and Germany, see Kevles, *In the Name of Eugenics;* Black, *War Against the Weak*.

4. See Wiener, *Charles S.Peirce: Selected Writings*.

5.This relation was dramatized by Kant through his development of the concepts of the *phenomena* of the world of all the things we normally observe and the *noumena* of all that lies beyond the limited reach of our sensory equipment and brain.

6.This became the primary language into which the scientific discourse on evolution was to be primarily locked for much of the 20th century. You could be respected as a historian, developmental psychologist, sociologist, or any other kind of scholar dabbling in speculation about our movement from past into future. But by and large to be classified as a properly credentialed, bona fide evolution theorist, you had to be to monumentally equipped to argue your case in terms of biology.

Evolutionary Systems Science, Political Science, and Economics

1. See note 8, chapter ten..

2. For Veblen, see Heilbronner, *The Worldly Philosophers.*

3. Boulding, *Evolutionary Economics.*

REFERENCES

Abraham, F., Abraham, *R.,* and Shaw, C. *A Visual Introduction to Dynamical Systems Theory for Psychology.* Aerial Press, 1990.

Abraham, F. An Historical Holistic Thread in the Dynamical Fabric of Psychology. In The Dynamics of Evolution: Essays in Honor of David Loye, Montuori, A., Ed., *World Futures: The Journal of General Evolution,* 49, 1-2, 159-201, 1997.

Abraham, R., and Shaw, C. *Dynamics: The Geometry of Behavior.* Addison-Wesley, 1992.

Adorno, T.W., Frenkel-Brunswick, E., Levinson, D.J., and Sanford, R. N. *The Authoritarian Personality.* Harper, 1950.

Anthony, M. *Integrated Intelligence.* Sense Publishers, 2008.

Barkow, J., Cosmides, L., and Tooby, J. *The Adapted Mind.* Oxford University Press, 1992.

Barrett, P., Weinshank, D.J., Ruhlen, P., and Ozminski, S. *A Concordance to Darwin's The Descent of Man.* Cornell University Press, 1987.

Bausch, K., and Christakis, A. Technology to Liberate Rather Than Imprison Consciousness. In Loye, D., Ed., *The Great Adventure: Toward a Fully Human Theory of Evolution.* SUNY Press, 2004.

Bausch, K. *The Emerging Consensus in Social Systems Theory.* Plenum, 2000.

Benedict, R. The Synergy Lectures. In Combs, A., Ed. *Cooperation: Beyond the Age of Competition.* Gordon and Breach, 1992.

Bible of the World. Viking,1939.

Black, E. *War Against the Weak: Eugenics and America's Campaign to Create a Master Race.* Dialogue Press, 2008.

Blumenthal, M. *Republican Gomorrah.* Nation Books, 2009.

Boulding, K. *Evolutionary Economics.* Sage, 1981.

Brown, A. *The Darwin Wars.* Simon & Schuster, 2001.

Burkhardt, F.. Ed. *Charles Darwin's Letters: A Selection, 1825-1859.*

Cambridge University Press, 1998.

Caldicott, H. *Nuclear Madness.* Norton, 1994.

Cantril, H. The Human Design. In Hollander, E.P., and R.G.Hunt, Eds., *Current Perspectives in Social Psychology.* Oxford University Press, 1967.

Capra, F. *The Web of Life.* Doubleday Anchor, 1996.

Carroll, S, and Loye, J. Natural Selection, Competition and Cooperation: Human Biology and Human Freedom. In Allan Combs, Ed., *Cooperation: Beyond the Age of Competition.* Gordon and Breach, 1964.

Chomsky, N., and Herman, E.S. *Manufacturing Consent.* Pantheon, 2002.

Clark, R.W. *The Survival of Charles Darwin.* Random House, 1984.

Colp, R. *To Be an Invalid: The Illness of Charles Darwin.* University of Chicago Press, 1977.

Combs, A., Ed. *Cooperation.* Taylor and Francis, 1992.

Combs, A. *The Radiance of Being.* Paragon House, 2002.

Csanyi, V. *Evolutionary Systems and Society.* Duke University Press, 1989.

Csikszentmihalyi, M. *The Evolving Self.* Harper,1993.

Dartmouth Bible. Chamberlin, R., and Feldman, H., Eds. Houghton-Mifflin, 1950

Darwin, C. *The Descent of Man.* Appleton,1879.

Darwin, C. *The Descent of Man.* Encylopedia Britannica Great Books, Vol.49, 1952.

Darwin, C.. *The Descent of Man.* Princeton University Press,1871/1981.

Darwin, C. *The Descent of Man.* In *Darwin Second Edition,* a CD-ROM published by Lightbinders, Inc., 1997.

Darwin, C. *The Descent of Man.* Penguin, 2004.

Darwin, C. *The Origin of Species.* In *Darwin Second Edition,* a CD-ROM published by Lightbinders, Inc., 1997. Also: Encyclopedia Britannica Great Books, Vol.49, 1952.

Darwin, C. *The Voyage of the Beagle.* In *Darwin Second Edition,* a CD-ROM published by Lightbinders, Inc., 1997. Also: P.F.Collier Harvard Classics,1909.

Darwin, C. *The Expression of the Emotions in Man and Animals.* In *Darwin Second Edition,* a CD-ROM published by Lightbinders, Inc., 1997. Also: University of Chicago Press, 1965.

Darwin, C. *Autobiography.* Norton,1887/1993.

Darwin, F. Reminiscences. In Darwin, C., *Autobiography.* Dover,

1892/1958.

Dawkins, R. *The Selfish Gene.* Oxford University Press, 1976.

Dawkins, R. *The Blind Watchmaker.* Norton, 1987.

de Chardin, T. 1955. *The Phenomonon of Man.* Fontana, 1955.

Depew, D., and Weber, B. *Darwinism Evolving.* MIT Press,1996.

Desmond, A., and Moore, J. *Darwin: The Life of a Tormented Evolutionist.* Penguin, 1991.

Desmond, A., and Moore, J. *Darwin's Sacred Cause: How a Hatred of Slavery Shaped Darwin's Views on Human Evolution.* Houghton Mifflin Harcourt, 2009.

de Quincey, C. *Radical Nature: The Soul of Matter.* Park Street Press, 2010.

Dobzhansky, T. Cultural Evolution, in D.L.Sills, Ed., *International Encyclopedia of the Social Sciences*, Vol.5. The Macmillan Company and the Free Press, 1968.

Dobzhansky, T. *Mankind Evolving.* Yale University Press,1987.

Dowd, M. *Thank GOD for Evolution!* Council Oak Books, 2007.

Durant, W., and Durant, A. *The Story of Civilization.* Historic eleven-volume set of books. Simon & Schuster, 1960s-1980s.

Eisler, R. *The Chalice and the Blade.* Harper & Row, 1987.

Eisler, R. *Sacred Pleasure.* HarperSanFrancisco, 1995.

Eisler, R. Action Research and Human Evolution: David Loye's Lifelong Exploration of Moral Sensitivity. *World Futures: The Journal of General Evolution* : 49, 1-2, 89-101, 1997.

Eisler, R. *Tomorrow's Children.* Westview Press, 2000.

Eisler, R. *The Power of Partnership.* New World Library, 2001.

Eisler, R., and Loye, D. *The Partnership Way.* Holistic Education Press, 1998.

Eisler, R. *The Real Wealth of Nations.* Berrett-Koehler, 2007.

Eldredge, N., and Gould, S. J. Punctuated equilibria: An alternative to phyletic gradualism. In Peter Schopf (Ed.) *Models in Paleobiology.* Freeman, Cooper, 1972.

Evans, R.I. *The Third Reich at War.* Penguin, 2009.

Forrest, B. *Understanding the Intelligent Design Creationist Movement.* Center for Inquiry, 2007.

Fox, M. *Original Blessing.* Tarcher, 2000..

Fuster, J. The Prefrontal Cortex. Lippincott-Raven, 1997.

Fromm, E. *Man for Himself: An Inquiry into the Psychology of Ethics.* Holt, Rinehart, and Winston, 1947.

Galtung, J., and Inayatullah, S., Eds. *Macrohistory and Macrohistorians.* Praeger, 1998.

Gilgoff, D. Gallup 'Darwin's Birthday' Poll: Fewer than Four in Ten Believe in Evolution. *U.S.News and World Report,* February 11, 2009.

Gleick, J. *Chaos.* Penguin, 1988.

Goerner, S. *Chaos and the Evolving Ecological Universe.* Routledge, 1994.

Goldie, P. *Darwin 2nd Edition.* A CD-ROM containing major books and papers by Darwin produced by Lightbinders, Inc., 1997.

Gorney, R. *The Human Agenda.* Simon and Schuster, 1972.

Gould, S.J. *Ever Since Darwin.* Norton, 1980.

Gould, S.J. *Bully for Brontosaurus.* Norton, 1991.

Gould, S.J. *The Structure of Evolution Theory.* Harvard University Press, 2002.

Greene, J. *The Death of Adam.* Iowa State University Press, 1959.

Gruber, H.E., and Barrett, P.H. *Darwin on Man.* Dutton, 1974.

Hamilton, W.D. The Evolution of Altruistic Behavior. *American Naturalist* (1963): 97:354-56.

Heilbroner, R. *The Worldly Philosophers.* Simon & Schuster, 1967.

Henderson, H., Houston, J., and Hubbard, B.M. *The Power of Yin.* Cosimo Books, 2007.

Hertsgaard, M. Who Killed Petra Kelly? *Mother Jones,* January/February, 1993.

Ho, M.W. Organism and Psyche in a Participatory Universe. In Loye, D., Ed., *The Evolutionary Outrider.* Praeger, 1998.

Hofstader, R. *Social Darwinism in American Thought.* Beacon, 1955.

Houston, J. *The Possible Human.* Tarcher, 1997.

Hubbard, B.M. *Conscious Evolution.* New World Library, 1998.

Huxley, J. *Evolutionary Ethics.* Oxford University Press, 1943.

Huxley, J. *Touchstone for Ethics 1893-1943.* Harper, 1947.

James, W. *Diary* MS p.55, James Papers, Houghton Library, Harvard University

James, W. Bain and Renouvier. *Nation* 22, 1876.

James, W. *The Principles of Psychology.* Dover, 1890/1950.

Jantsch, E. *The Self-Organizing Universe.* Pergamon Press, 1980.

Jay, M. *The Dialectical Imagination.* Little-Brown, 1973.

Jianin, Min, Ed. *The Chalice and the Blade in Chinese Culture.* China Social Sciences Publishing House, 1995.

Kant, I. *The Critique of Practical Reason.* New York: Macmillan, 1993.

Kant, I. *Groundwork of the Metaphysics of Morals.* Cambridge University Press, 1998.

Kauffman, S.A. *At Home in the Universe.* Oxford University Press, 1996.

Kauffman, S.A. *Reinventing the Sacred.* Basic Books, 2010.

Kennedy, D.J. *Darwin's Deadly Legacy* (DVD). Corel Ridge Ministries, 2006

Kevles, D. *In the Name of Eugenics.* Harvard University Press, 1998.

Keynes, R. *Darwin, His Daughter, and Human Evolution.* Riverhead, 2001.

Krippner, S. *Human Possibilities.* Doubleday Anchor, 1980

Krippner, S. Running with Spotted Fawn in the Akashic Experience. In Laszlo, E., Ed, *The Akashic Experience.* Inner Traditions, 2009

Kropotkin, P. *Mutual Aid: A Factor of Evolution.* Porter Sargent, 1950.

Kropotkin, P. *Ethics: Origins and Development.* Dial Press, 1924.

Kung, H., and Kuschel, K.J., Eds. *A Global Ethic: The Declaration of the Parliament of the World's Religions.* SCM Press, 1993.

Kung, H. *A Global Ethic for Global Politics and Economics.* SCM Press, 1997.

Kung, H. The Beginning of All Things: Science and Religion. Eerdmans, 2008

Kuhn, T. *The Structure of Scientific Revolution.* University of Chicago Press, 1970.

Lakoff, G. *Moral Politics.* University of Chicago Press, 2002.

Lakoff, G. *The Political Mind.* Viking, 2008.

Laszlo, E., Ed. *The New Evolutionary Paradigm.* Taylor and Francis, 1991.

Laszlo, E. *The Choice: Oblivion or Evolution.* Tarcher, 1994.

Laszlo, E. *Evolution: The General Theory.* Hampton Press, 1996.

Laszlo, E. *Science and the Akashic Field: An Integral Theory of Everything.* Inner Traditions, 2007.

Laszlo, E., Ed. *The Akashic Experience.* Inner Traditions, 2009.

Lerner, M. *The Left Hand of God.* Harper One, 2007.

Lerner, M. *The Politics of Meaning.* Perseus Books, 1997.

Lewin, K. *Resolving Social Conflicts.* Harper, 1948.

Lewin, K. *Field Theory in Social Science.* Harper, 1951.

199

Lewontin, R.C., Rose, S. and Kamin, L. *Not in Our Genes*. Pantheon, 1984.

Lorenz, E. Irregularity: A Fundamental Property of the Atmosphere. *Tellus*, 36A, pp.98-110.

Loye, D. *The Healing of a Nation*. Norton, 1971.

Loye, D., and Rokeach, M. Ideology, Belief Systems, Values, and Attitudes. In *International Encyclopedia of Neurology, Psychiatry, Psychoanalysis and Psychology*. Van Nostrand, 1976.

Loye, D., and Eisler, R. Chaos and Transformation: The Implications of Natural Scientific Nonequilibrium Theory for Social Science and Society. *Behavioral Science* (1987): 32, 1, pp.53-65.

Loye, D. Moral Sensitivity and the Evolution of Higher Mind. In Laszlo, E., Masuli, I., Artigiani, R., and Csanyi, V., Eds. *The Evolution of Cognitive Maps*. Gordon and Breach, 1993.

Loye, D. How Predictable is the Future: The Conflict Between Traditional Chaos Theory and the Psychology of Prediction, and the Challenge for Chaos Psychology." In Robin Robertson and Allan Combs, Eds., *Chaos Theory in Psychology and the Life Sciences*. Erlebaum, 1995.

Loye, D., Ed. *The Evolutionary Outrider: The Impact of the Human Agent on Evolution*. Praeger, 1998.

Loye, D. Evolutionary Action Theory: A Brief Outline. In Loye, D., Ed., *The Evolutionary Outrider*. Praeger, 1998.

Loye, D. The Moral Brain. *Brain and Mind 3* (2002): 133-150.

Loye, D., Ed. *The Great Adventure: Toward a Fully Human Theory of Evolution*. SUNY Press, 2004.

Loye, D. *Bankrolling Evolution*. Benjamin Franklin Press, 2007

Loye, D. *Measuring Evolution*. Benjamin Franklin Press, 2007.

Loye, D. *Darwin in Love*. Benjamin Franklin Press, 2010.

Loye, D. *Darwin's Lost Theory, 2nd edition*. Benjamin Franklin Press, 2010.

Luria, A.. *The Working Brain*. Basic Books, 1973.

MacLean, P. *The Triune Brain in Evolution: Role in Paleocerebral Functions*. Plenum Press, 1990.

Margulis, L., and Sagan, D. *Origins of Sex: Three Billion Years of Genetic Recombination*. Yale University Press, 1986.

Margulis, L., and Sagan, D. *The Mystery Dance: On the Evolution of Human Sexuality*. Summit Books, 1991.

Marx, K. *Capital: A Critique of Political Economy*. Penguin, 1992.

200

Marrow, A. *The Practical Theorist: The Life and Work of Kurt Lewin.* Basic Books, 1969.

Maslow, A. *Toward a Psychology of Being.* Van Nostrand, 1968.

Maslow, A. The Psychology of Science. Regnery, 1970.

Maslow, A. *The Farther Reaches of Human Nature.* Viking, 1971.

Miller, G.A., Gallanter, A., and Pribram, K.H. *Plans and the Structure of Behavior.* Holt, 1960.

Monod, J. BBC Interview, July 1970, quoted in *Beyond Chance and Necessity*, John Lewis, Ed. Teilhard Centre for the Future, 1974.

Monod, J. *Chance and Necessity.* Vintage, 1971.

Montagu, A. *The Direction of Human Development.* Hawthorn, 1970.

Montagu, A. *The Nature of Human Aggression.* Oxford University Press, 1976.

Moore, J.M. Socializing Darwin. In L.Levidow, Ed., *Science as Politics* Free Association Press, 1986.

Mrydal, Gunnar. *Objectivity in Social Research.* Wesleyan, 1988.

Neibuhr, R. *The Nature and Destiny of Man.* Westminster John Knox Press, 1996.

Noddings, N. Caring: *A Feminine Approach to Ethics and Moral Education.* University of California Press, 2003.

Nuclear Weapons: Report of the Secretary-General. Autumn Press, 1980.

Nussbaum, P. Evangelicals Divided Over Evolution, *Philadelphia Inquirer,* June 4, 2005.

Obama, B. *Dreams from My Father.* Three Rivers Press, 2004.

O'Manique, J. *The Origins of Justice.* University of Pennsylvania Press, 2002.

Ornstein, R. *The Evolution of Consciousness.* Prentice Hall, 1991.

Peake, A. *The Daemon.* Arcturus, 2008.

Piaget, J. *The Moral Judgement of the Child.* Free Press, 1965.

Pribram, K. *Brain and Perception.* Erlebaum, 1991.

Pribram, K. On Brain, Conscious Experience, and Human Agency. In Loye, D., Ed., *The Evolutionary Outrider.* Praeger, 1998.

Prigogine, I., and Stengers, I. *Order Out of Chaos.* Bantam, 1984.

Prindle, D.F. *Stephen Jay Gould and the Politics of Evolution.* Prometheus Books, 2009.

Raffi, *Child Honoring.* Homeland Press, 2006.

Robbins, J. *The Food Revolution.* Conari Press, 2001.

Richards, R.J. *Darwin and the Emergence of Evolutionary Theories of Mind and Behavior.* University of Chicago Press, 1987.

Richards, R., Ed. *Everyday Creativity and New Views of Human Nature.* American Psychological Association, 2007.

Richerson, P., and Boyd, R. *Not by Genes Alone.* University of Chicago Press, 2004.

Rokeach, M. *The Nature of Human Values.* Free Press. 1973.

Romanes, Ethel. *Life and Letters of George John Romanes.* Longmans, Green, 1896.

Romanes, G. *Darwin and After Darwin.* Longmans Green, 1893.

Salthe, S. *Development and Evolution.* MIT Press, 1996.

Schneider, K., Bugenthal, J., and Pierson, F., Eds. *The Handbook of Humanistic Psychology.* Sage, 2002.

Schlesinger, A., Sr. The Tides of American Politics, in *Paths to the Present.* Houghton Mifflin, 1964.

Smith, A. *The Theory of Moral Sentiments.* Clarendon Press, 1976.

Smith, C., and Beccaoni, G. *Natural Selection and Beyond: The Intellectual Legacy of Alfred Russel Wallace.* Kindle Books, 2009.

Sober, E., and Wilson, D.S. *Unto Others: The Evolution and Psychology of Unselfish Behavior.* Harvard University Press, 1998.

Swimme, B., and Berry, T. *The Universe Story.* HarperSanFrancisco, 1992.

Tarnas, R. *The Passion of the Western Mind.* Ballantine, 1993.

Toffler, A. *Future Shock.* Random House, 1970.

Trivers, R. *Social Evolution.* Benjamin-Cummings, 1985.

Varela, F. *Ethical Know-How.* Stanford University Press, 1999.

Wallace, A.R. *A Defense of Modern Spiritualism.* Colby and Rich, 1874.

Weber, B., and Depew, D. *Evolution and Learning: The Baldwin Effect Reconsidered.* MIT Press, 2007.

Weikart, Richard. *From Darwin to Hitler.* Palgrave-Macmillan, 2006.

Weiner, N. *Cybernetics: Or the Control and Communication in the Animal and the Machine.* MIT Press, 1985.

Weinshank, D.J., Osminski, S., Ruhlen, P., and Barrett, W. *A Concordance to Charles Darwin's Notebooks, 1836-1844.* Cornell University Press, 1989.

Wells, H.G. *The Time Machine.* Penguin, 2005.

Wesson, R. *Beyond Natural Selection.* MIT Press, 1991.

Wiener, P.. Editor. *Values in a Universe of Chance:* Selected Writings of

Charles S. Peirce. Stanford University Press, 1958.

Wilber, K. *The Atman Project.* Quest Books, 1980.

Wilber, K. *Integral Psychology: Consciousness, Spirit, Psychology, Therapy.* Shambhala, 2000.

Wilber, K. *A Brief History of Everything.* Shambhala, 2007.

Wilson, D.S. and Sober, E. Reintroducing Group Selection to the Human Behavioral Sciences. *Behavioral and Brain Sciences* (1994): 17, 585-608.

Wilson, D.S. *Darwin's Cathedral.* University of Chicago Press, 2003.

Wilson, D.S. *Evolution for Everyone.* Delta, 2007.

Wilson, E.O. *Sociobiology.* Harvard University Press, 1975.

Wilson, E.O. *Human Nature.* Harvard University Press,1978.

Wilson, E.O. *The Diversity of Life*. Harvard University Press, 1992.

Wink, W. The Powers That Be: Theology for a New Millennium. Gallilee Trade, 1999.

Zimmerman, M. Combating the Fifth Wave of Creationism: Religious Leaders and Scientists Working Together. *Theology and Science,* 8:2, 211-222, 2010.

Zimmerman, M., and Loye, D. The New Wave of Creationism. *World Futures: The Journal of General Evolution* (in press).

INDEXES

Name Index
Subject Index
Darwin Index

NAME INDEX

207

SUBJECT INDEX

211

DARWIN INDEX

This index is laid out by category, and in sub-ordering of items, to facilitate development of updated curricula for the teaching of the second and completing half as well as the introductory first half for Darwin's theory and the story of human evolution.

Darwin and

people
Romanes, 5, 9, 11, 12
Galton, 24, 43
Maslow, 18
Wallace, 23, 41
T.H.Huxley, 49, 89
James Mackintosh, 105
Immanuel Kant, 105
son Leonard, 47

major books
The Origin of Species, 17, 18, 119, 120
The Descent of Man, 16, 18, 83, 159

theory building
completion of theory, 4, 9, 130, 154
first revolution, 4
second revolution, 4, 15, 78, 153, 154, 163

first half theory
competition, 3, 114, 167

gradual evolution, 50
natural selection, 3, 10, 13-15, 17, 19, 21-25, 29, 31, 40, 41, 44-46, 48-50, 72, 76, 87, 91, 96, 107, 127, 128, 130, 131, 133, 148-150
selfishness, 2, 14, 52, 82, 84-86, 88, 92, 121, 130, 150, 162, 167
survival of the fittest, 2, 49, 52, 60, 71, 72, 84, 87, 91, 106, 121, 130, 146, 150, 161
variation, 14, 17, 18, 21, 40, 41, 44-46, 48, 50, 72, 76, 91, 107, 132, 148-150

completion of theory
altruism, 2, 85-88, 92
brain, 3, 56, 83
capacity for emotion, 16
capacity for reason, 16
caring, 2, 16, 17, 69, 71, 121, 139, 140, 162
community selection, 25
cooperation, 60, 87, 93, 114, 130, 167
correlated variation, 18, 132
cultural evolution, 17
education, 4, 15, 28, 47, 56, 57, 70, 81, 89, 96, 117, 118, 128, 137, 139, 140, 155, 161, 163, 169
group selection, 25
habit, 3, 17
instruction, 3, 98
language, 17, 53, 55, 59, 61-64, 66,

CPSIA information can be obtained at www.ICGtesting.com
Printed in the USA
BVOW031045290612

294014BV00010B/127/P

9 780979 525759